Discovering Structural Equation Modeling Using Stata

Revised Edition

D0713966

Discovering Structural Equation Modeling Using Stata

Revised Edition

ALAN C. ACOCK
Oregon State University

A Stata Press Publication
StataCorp LP
College Station, Texas

Published by Stata Press, 4905 Lakeway Drive, College Station, Texas 77845
Typeset in LaTeX 2_ε
Printed in the United States of America
10 9 8 7 6 5 4 3 2 1

ISBN-10: 1-59718-139-0
ISBN-13: 978-1-59718-139-6

Library of Congress Control Number: 2013946346

Dedication

This book is dedicated to the graduate students and professionals who have participated in my seminars and workshops on structural equation modeling.

Contents

Tables

Figures

Boxes

Preface

What is assumed?

There are two ways of learning about structural equation modeling (SEM). The one I have chosen for this book is best described by an old advertising tag for a sport shoe company: "Just do it". My approach could be called kinetic learning because it is based on the tactile experience of learning about SEM by using Stata to estimate and interpret models. This means you should have Stata open while you read this book; otherwise, this book might help you go to sleep if you try to read it without simultaneously working through it on your computer. By contrast, if you do work through the examples in the book by running the commands as you are reading, I hope you develop the same excitement that I have for SEM.

The alternative approach to learning SEM is to read books that are much more theoretical and may not even illustrate the mechanics of estimating models. These kinds of books are important, and reading them will enrich your understanding of SEM. This book is not meant to replace those books, but simply to get you started. My intent is for you to work your way through this book sequentially, but I recognize that some readers will want to skip around. I am hopeful that after you have been through this book once, you will want to return to specific chapters to reference techniques covered there. To facilitate this, each chapter includes some repetition of the most salient concepts covered in prior chapters. There is also a detailed index at the end of the book.

What background is assumed? A person who has never used Stata will need some help getting started. A big part of Stata's brilliance is its simplicity, so a few minutes of help will get you up and ready for what you need to know about Stata. If you are new to Stata, have a friend who is familiar with the program show you the basics. If you have read my book *A Gentle Introduction to Stata* (2012a), you are ahead of the game. If you have any experience using Stata, then you are in great shape for this book. If you are a longtime Stata user, you will find that parts of this book explain things you already know.

To get the most out of this book, you need to have some background in statistics with experience in multiple regression. If you know path analysis, you will find the SEM approach to path analysis a big improvement over traditional approaches; however, the material on path analysis has been written for someone who has had very little exposure to path analysis. Even though the first chapter begins by covering how factor analysis has been used traditionally, a background in factor analysis is less important than having had some exposure to multiple regression. The first chapter shows how confirmatory factor analysis adds capabilities to move beyond the traditional approach—you may never want to rely on alpha and principal component factor analysis again for developing a scale. I have covered enough about the traditional applications of factor analysis that you will be okay if you have had little or no prior exposure to factor analysis.

What will I learn?

We will explore many of the most widely used applications of SEM. We will begin with how to estimate a confirmatory factor analysis model—this is the measurement model part of SEM. This chapter includes parceling as a way to handle a large number of items for one or more of your factors. Next, we will cover path models using SEM—this is the structural model part of SEM. This chapter also introduces nonrecursive path models. We then put these two components together to introduce the full structural equation model. This chapter on the full model includes a number of specialized actions, such as equality constraints. With this foundation, we move on to a chapter on growth curves and conclude with a chapter on multiple-group analysis.

The book has two appendixes. Appendix A shows you how to use Stata's graphical user interface (GUI) to draw and estimate models with Stata's SEM Builder. It would be very useful to begin here so that you are familiar with the SEM Builder interface. If you have no background in SEM, you will not understand how to interpret the results you generate in appendix A, but this is not the point. Appendix A is just there to acquaint you with the SEM Builder that Stata introduced in version 12 and enhanced in version 13. How the interface works is the focus of appendix A. In the text, I use this GUI fairly often, but the focus is on understanding why we are estimating models the way we do and how we interpret and present the results. All the figures presented in this book were created using the SEM Builder, which produces publication-quality figures—far better than what you can draw with most other software packages that produce "near" publication-quality figures.

Appendix B shows you how to work with summary data (means, standard deviations, correlations) that are often reported in published works. You will be able to fit most models with these summary statistics even if you do not have the real data. This feature is great when you read an article and would like to explore how alternative models might be more appropriate. Many articles include a correlation matrix along with standard deviations and means. If these are not included, it is easier to request them from the author than it is to request the author's actual data.

In addition to the two main appendixes, the first two chapters each have their own appendix that briefly describes using the SEM Builder for the models estimated in that chapter.

How do I use this book?

The chapters are intended to be read in the order they appear, and you should follow along with your reading on your computer. (It does not matter whether you use a Windows, Mac, or Unix operating system because the Stata commands and results will be identical.) If you have no background in SEM or factor analysis, take your time reading chapter 1. If you are comfortable with SEM and factor analysis, you should still go over chapter 1 enough to get a feel for how Stata works with the measurement model.

Though the chapters are fairly long, they are broken up into more manageable sections. If you are like me, once you know the commands I cover, you will have enough on your plate that you will forget the specifics before you need to fit a particular type of model. The sections in each chapter build on each other but are sufficiently independent that you should find them useful as a reference. Someday you will want to estimate a nonrecursive path model or a mediation model; you can easily find the section covering the appropriate model and come back to it. At the same time, this book does not attempt to compete with Stata's own *Structural Equation Modeling Reference Manual*, or [SEM]; I only cover a widely used subset of the options and postestimation commands available in Stata's SEM package.

What resources are available?

To facilitate the kinetic part of learning, you can download all the data used in this book as well as the Stata programs, called do-files, that fit every model. In the Command window, type the following:

```
. net from http://www.stata-press.com/data/dsemusr/
. net describe dsemusr
. net get dsemusr
```

When you run these three commands, you do not type the initial period and space, called the dot prompt. A convention in all Stata documentation and output in the Results window is to include the dot prompt as a prefix to each command, but you need only type the command itself.

There are several varieties of Stata software, and all of these are able to run the models described in this text. I focus on the Windows and Mac operating systems, and I show when there are slight differences in how they work in the GUI. The Unix GUI is very similar to the Windows GUI. The same Stata do-files run on all operating systems, though the systems differ slightly in how the file structure is organized.

One variety of Stata is called Small Stata. This is full featured and is small only in the sense of being limited in the number of observations (1,200) and variables (99) it can handle. Because a few of the datasets I use have more than 1,200 observations, I have made up smaller datasets that will work using Small Stata. You can obtain these datasets by entering the following in the Command window:

```
. net from http://www.stata-press.com/data/dsemusr/
. net describe dsemusr_small
. net get dsemusr_small
```

Using the Small Stata data, you will get somewhat different results for some models in the book simply because you will be using a smaller dataset. In addition, there are three models in chapter 4 that will not run using Small Stata.

At the end of each chapter, you will find some exercises that illustrate the material covered in the chapter. It is important to fit all the models in the text while you read the book because this reinforces what you are learning, as does typing in the commands yourself. The exercises extend this learning process by having you develop your own set of commands and models using the GUI system.

There is much more to SEM than could possibly be covered in a book this size. This book is intended to complement the material in the Stata manuals (over 11,000 pages of helpful information), which are available as PDF files when you install Stata. One way to access the [SEM] manual is to type `help sem` in the Command window of Stata. This opens a help file. At the top left of the help file, the title ([SEM] `sem and gsem`) is highlighted in blue. Clicking on this blue link will open up the PDF file of the [SEM] manual.

This book contains a fairly detailed index. Although I have explanatory section headings and these are a good place to start searching for how to do something, the index is naturally much more detailed. You may need to find how to place equality constraints on a multiple-group analysis or on a pair of reciprocal paths. These are covered in very different sections of the book, and the index tells you where to find them. The index was written to be useful after you have read this book and are using it as a reference to guide you while fitting your own models on your own data.

Conventions

Typewriter font. I use a `typewriter font` when something would be meaningful to Stata as input. This would be the case for something you type in the Command window or in a do-file. If a command is separated from the main text, as in

```
. sem (compliance <- educ income gender)
```

a dot prompt will precede the command. I also use a typewriter font for all Stata results, variable names, folders, and filenames.

Bold font. I use a **bold font** for menu items and for buttons you click within a menu. The bold font helps distinguish the button from the text; for example, you might be instructed to click the **Adjust Canvas Size** button.

Slant font. I use a *slant font* when referring to keys on our keyboard, such as the *Enter* key.

Italic font. I use an *italic font* when referring to text in a menu that you need to replace with something else, such as an actual variable name.

Capitalization. Stata is case sensitive. The command `sem (compliance <- educ income gender)` will produce a maximum likelihood multiple regression. If you replace `sem` with `Sem`, Stata will report that it has no command called `Sem`. I will use lowercase for all commands and all observed variables. When I refer to latent variables, I will capitalize the first letter of the latent variable. A simple confirmatory factor analysis would be `sem (Alienation -> anomia isolate depress report)`. Only the latent variable, `Alienation`, is capitalized. The arrow indicates that observed variables measure how a person responds on an anomia scale labeled `anomia`, an isolation scale labeled `isolate`, a depression scale labeled `depress`, and a reported score from an observer labeled `report`. All four of these observed variables depend on their level of `Alienation`, the latent variable.

Acknowledgments

I acknowledge the extraordinary support of the Stata staff who have worked with me on this project. Kristin MacDonald was the statistical reviewer of this book, but her contribution is much greater than you might expect from a reviewer. Because this book was written for the first version of the structural equation modeling capabilities in Stata, I was learning new options as I was writing the book. Kristin MacDonald pointed out better ways of doing many of the procedures covered in this book. Her advice has greatly enhanced the value of this book. Writing a book on an advanced statistical application with the intention of making it accessible to people who are just learning the application is challenging. Deirdre Skaggs was the English editor, and she made innumerable changes in the wording that helped to clarify what I was trying to write. Annette Fett designed the cover. The idea of learning a new application and creating a new research capability was cleverly captured in her design. Kaycee Headley assisted me in preparing a draft for reviews. I want to thank Stephanie White, who converted my text to proper LaTeX, which is the document markup language used for TeX typesetting. Those of you who are familiar with LaTeX can fully appreciate what this means. Stata works very well with LaTeX, and all the tables of results that appear in the book are exactly what you see when you run Stata.

Finally, I want to thank the leadership of Stata Press. I know of no other publisher who even approaches the level of author support they provide. To the extent this book lives up to the excellence of Stata Press publications, much of the credit goes to the leadership.

1 Introduction to confirmatory factor analysis

1.1 Introduction

When we are measuring a concept, it is desirable for that concept to be unidimensional. For example, if we are measuring a person's perception of his or her health, the concept is vague in that there are multiple dimensions of health—physical health, mental health, and so on. Let us suppose Chloe is in excellent physical health (a perfect 10), but she is very low on mental health (a score of 0.0). Madison is in excellent mental health (a perfect 10), but has serious problems with her physical health (a score of 0.0). Jaylen is average on both dimensions (a score of 5.0 on both). Do we want to give all three the same score because Chloe, Madison, and Jaylen each average 5.0? Should one dimension be more important than another?

When you have two dimensions (x and y) and try to represent them on a graph, you need two values: one showing a score on the x dimension and one showing a score on the y dimension. When there is more than one dimension, a single score becomes difficult to interpret, and it is often misleading to represent the location of a person on the concept with a single number. Thus there are advantages to narrowly defining our concepts so our measures can tap a single dimension. If we are interested in multiple dimensions, such as distinguishing between physical and mental health, then we need multiple concepts and multiple empirical sets of measures.

On the other hand, we can carry this argument too far. Each item we might pick to measure physical health will represent a slightly different aspect of physical health. We should aim to represent as broad a meaning of physical health as we can without adding distinctly different dimensions. The ideal way to do this is to allow each item to have its own unique variance and develop a scale score that represents the shared meaning of the set of items on a single dimension. This way, our measurement model represents concepts that are neither too broad to have a clear meaning nor too narrow to be of general interest.

This is where we will go with confirmatory factor analysis (CFA). We will first cover the "do not even think about it" approach, followed by the exploratory search for a single dimension using the traditional principal component factor analysis approach. We will then extend this to CFA measurement models where we have first one and then two or more concepts.

1.2   The "do not even think about it" approach

Many studies have brief sections reporting how they measured the variables. The authors simply assume the dimensionality of what they are measuring; all they report is the alpha (α) measure of reliability. (Do not confuse this alpha coefficient with the use of alpha for the conventional level of statistical significance.) If alpha is greater than 0.80 (sometimes a lower value is considered adequate), then these authors say they have a good measure. The alpha coefficient is a measure of internal consistency. It depends on just two parameters, namely, the average correlation/covariance of the items with one another and the number of items. With 20 or more items, the alpha could be 0.80 even if the items are only weakly correlated with one another and even if the items represent several dimensions.

A good alpha value does not ensure that a single dimension is being tapped. Consider the following correlation matrix:

	$x1$	$x2$	$x3$	$x4$	$x5$	$x6$
$x1$	1.0					
$x2$	0.6	1.0				
$x3$	0.6	0.6	1.0			
$x4$	0.3	0.3	0.3	1.0		
$x5$	0.3	0.3	0.3	0.6	1.0	
$x6$	0.3	0.3	0.3	0.6	0.6	1.0

We see in this matrix two subsets of items: $x1$–$x3$ and $x4$–$x6$. Items $x1$–$x3$ are all highly correlated with each other (r's $= 0.6$) but much less correlated with items $x4$–$x6$ (r's $= 0.3$). Similarly, items $x4$–$x6$ are highly correlated with one another (r's $= 0.6$) but much less correlated with items $x1$–$x3$ (r's $= 0.3$). This indicates that there are two related dimensions, namely, whatever is being measured by $x1$–$x3$ for one dimension and whatever is being measured by $x4$–$x6$ for the other. The alpha for these six items is $\alpha = 0.81$, which is considered good. For example, Kline (2000) indicates that an alpha of 0.70 and above is acceptable. However, the point here is that when we rely on alpha to justify computing a total or mean score for a set of items, we may be forcing together two or more dimensions, that is, trying to represent two (or more) concepts with one number. At the very least, we should routinely combine reports of reliability with some sort of factor analysis to evaluate how many dimensions we are measuring.

Alpha can be high even with items that are only minimally related to one another. The formula for a standardized alpha is

$$\alpha = \frac{k\bar{r}}{1 + (k - 1)\bar{r}}$$

where k is the number of items in the scale and $\bar{r}$ is the mean correlation among the items. We would not think of an $r = 0.17$, for example, as more than a minimal relationship. After all, if $r = 0.17$ then $r^2 = 0.03$, meaning that 97% of the variance in the two variables is not linearly related. However, if you had a 40-item scale with an average correlation of just 0.17, your alpha would be 0.80. The measure would be reliable in the sense of internal consistency, but the high alpha does not mean we are measuring a single dimension. Simply adding up a series of items (or taking their mean) and reporting an alpha is insufficient to qualify a measure as a good measure.

1.3 The principal component factor analysis approach

Principal component factor analysis (PCFA) is the most traditional approach to factor analysis. Several authors have demonstrated that this is far from the best type of factor analysis (Fabrigar et al. 1999; Costello and Osborne 2005), and some prefer to go so far as to say it is not really factor analysis at all. Stata offers alternative exploratory factor analysis methods, including maximum likelihood factor analysis, that have significant advantages; we are using Stata's PCFA only because of its widespread use. The structural equation modeling approach has advantages over all the traditional approaches to factor analysis and will be the focus of this book.

A major concern with PCFA is that it tries to account for all the variance and covariance of the set of items rather than the portion of the covariance that the items share in common. Thus it assumes there is no unique or error variance in each of the indicator variables. One reason PCFA is so widely used is because it is the default method in other widely used statistical packages, and you need to override this default in those programs to get a truer form of factor analysis. In Stata, PCFA is an option you need to specify and not the default. The Stata command for PCFA is simply `factor` *varlist*`, pcf`, where `pcf` stands for principal component factor analysis. Through the menu system, click on **Statistics > Multivariate analysis > Factor and principal component analysis > Factor analysis**.[1] In that dialog box, you list your variables under the **Model** tab. Under the **Model 2** tab, you pick *Principal-component factor*.

We will illustrate PCFA using actual data from the National Longitudinal Survey of Youth, 1997 (NLSY97). This is a longitudinal study that focuses on the transition from youth to adulthood. In 2006, when the participants were in their 20s, the NLSY97 asked a series of questions about the government being proactive in promoting well-being. The questions covered such topics as providing decent housing, college aid, reducing the income differential, health care, and providing jobs. We are interested in using 10 items to create a measure of conservatism. In the `nlsy97cfa.dta` dataset, these items are named `s8646900`–`s8647800`; for simplicity, we have renamed them `x1` to `x10`. The commands appear in a do-file called `ch1.do`, which you can find at http://www.stata-press.com/data/dsemusr/ch1.do. The dataset is located at

```
. use http://www.stata-press.com/data/dsemusr/nlsy97cfa.dta
```

---

1. Warning: When you go to **Statistics > Multivariate analysis > Factor and principal component analysis**, do not then pick **Principal component analysis (PCA)** from the menu. This is intended to extract principal components, linear combinations of the variables, rather than factors.

Before constructing a scale, we need to examine the items. As you can see below, the responses range from 1 (meaning the task definitely should be the role of government) to 4 (meaning the task definitely should not be the role of government). Higher scores indicate that the person is more conservative.

```
. codebook x1-x10, compact
Variable    Obs Unique     Mean  Min  Max  Label

x1         1833      4  2.331697    1    4  GOVT RESPONSIBILITY - PROVIDE JOB...
x2         1859      4  1.620226    1    4  GOVT RESPNSBLTY - KEEP PRICES UND...
x3         1874      4  1.416222    1    4  GOVT RESPNSBLTY - HLTH CARE FOR S...
x4         1872      4  1.365385    1    4  GOVT RESPNSBLTY -PROV ELD LIV STA...
x5         1815      4  1.773003    1    4  GOVT RESPNSBLTY -PROV IND HELP 2006
x6         1811      4  2.276643    1    4  GOVT RESPNSBLTY -PROV UNEMP LIV S...
x7         1775      4  2.228732    1    4  GOVT RESPNSBLTY -REDUCE INC DIFF ...
x8         1875      4  1.309333    1    4  GOVT RESPNSBLTY -PROV COLL FIN AI...
x9         1847      4  1.705468    1    4  GOVT RESPNSBLTY -PROV DECENT HOUS...
x10        1860      4   1.39086    1    4  GOVT RESPNSBLTY -PROTECT ENVIRONM...
```

A PCFA can be run on these items by using a very simple command:

```
. factor x1-x10, pcf
(obs=1617)

Factor analysis/correlation                   Number of obs    =     1617
    Method: principal-component factors       Retained factors =        2
    Rotation: (unrotated)                     Number of params =       19
```

Factor	Eigenvalue	Difference	Proportion	Cumulative
Factor1	3.91523	2.90094	0.3915	0.3915
Factor2	1.01014	0.13285	0.1014	0.4930
Factor3	0.88144	0.11496	0.0881	0.5811
Factor4	0.76648	0.02404	0.0766	0.6577
Factor5	0.74243	0.04889	0.0742	0.7320
Factor6	0.69354	0.08649	0.0694	0.8013
Factor7	0.60705	0.06820	0.0607	0.8620
Factor8	0.53886	0.09140	0.0539	0.9159
Factor9	0.44746	0.05424	0.0447	0.9607
Factor10	0.39322	.	0.0393	1.0000

```
LR test: independent vs. saturated:  chi2(45) = 4083.46 Prob>chi2 = 0.0000
Factor loadings (pattern matrix) and unique variances
```

Variable	Factor1	Factor2	Uniqueness
x1	0.6064	-0.3789	0.4888
x2	0.5810	0.0438	0.6605
x3	0.7221	0.2140	0.4328
x4	0.7174	0.3200	0.3830
x5	0.5780	-0.0261	0.6653
x6	0.6091	-0.4536	0.4233
x7	0.6050	-0.3327	0.5233
x8	0.5994	0.3252	0.5350
x9	0.7330	-0.1621	0.4365
x10	0.4543	0.5211	0.5221

These results indicate that the first factor is very strong with an eigenvalue of 3.92. The eigenvalue is how much of the total variance over all the items is explained by the first factor. By using the `display` command, we can compute the first eigenvalue as the sum of the squared factor loadings for the first factor:

```
. display .6064^2+.5810^2+.7221^2+.7174^2+.5780^2+.6091^2+.6050^2 +.5994^2+
> .7330^2+.4543^2
3.9154428
```

The PCFA analyzes the correlation matrix where each item is standardized to have a variance of 1.0. Therefore, with 10 items, the eigenvalues combined will add up to 10. With 3.92 out of 10 being explained by the first factor, we say the first factor explains 39.2% of the variance in the set of items. Any factor with an eigenvalue of less than 1.0 can usually be ignored.

The second factor has an eigenvalue of 1.01, which is very weak though it does not strictly fall below the 1.0 cutoff. We decide that the first factor, explaining 39.2% of the variance in the 10 items, is the only strong factor. This is reasonably consistent with our intention to pick items that tap a single dimension. We do not have an explicit test of a single-factor solution, but the eigenvalue of 3.92 is large enough to be reasonably confident that all the items are tapping a single dimension. Notice that all the loadings of the items of `Factor1` are substantial, varying from 0.45 to 0.73. This range is also good when compared to conventions of the loadings being 0.4 or above. Some authors feel a loading of at least 0.30 is the minimum criterion for an item (Costello and Osborne 2005). You may recall that with the PCFA, the loadings are the correlation between how people respond to each item and the underlying, latent dimension.

Even though the last item has a loading over 0.40, its loading is considerably weaker than the rest of the items. The last item is about the environment, which can be a personal concern of anyone, whether conservative or not. By contrast, the other nine items involve government response to needs people have because of their limited personal resources. Because there is a second factor with an eigenvalue greater than 1.0 and because the loading of the tenth item on the first factor is the weakest, we will drop that item and rerun our analysis to see if we can obtain a clearer result.

```
. factor x1-x9, pcf
(obs=1625)
```

Factor analysis/correlation   Number of obs   =   1625
    Method: principal-component factors   Retained factors =   1
    Rotation: (unrotated)   Number of params =   9

Factor	Eigenvalue	Difference	Proportion	Cumulative
Factor1	3.76124	2.80650	0.4179	0.4179
Factor2	0.95473	0.10627	0.1061	0.5240
Factor3	0.84847	0.10176	0.0943	0.6183
Factor4	0.74671	0.05561	0.0830	0.7012
Factor5	0.69110	0.07429	0.0768	0.7780
Factor6	0.61681	0.07780	0.0685	0.8466
Factor7	0.53900	0.09177	0.0599	0.9065
Factor8	0.44723	0.05252	0.0497	0.9561
Factor9	0.39471	.	0.0439	1.0000

LR test: independent vs. saturated: chi2(36) = 3863.18 Prob>chi2 = 0.0000

Factor loadings (pattern matrix) and unique variances

Variable	Factor1	Uniqueness
x1	0.6243	0.6103
x2	0.5883	0.6539
x3	0.7222	0.4785
x4	0.7131	0.4915
x5	0.5818	0.6615
x6	0.6197	0.6160
x7	0.6085	0.6297
x8	0.5968	0.6439
x9	0.7392	0.4535

Most researchers would be quite happy with these results. Only one factor has an eigenvalue greater than 1.0, and all nine items load over 0.5 on that factor.

Many researchers ignore the results shown in the column labeled `Uniqueness`. These values represent the unique variance or error variance. For example, 61% of the variance in indicator variable `x1` is not accounted for by the factor solution. The principal component factor method assumes that these uniquenesses are 0. The uniquenesses are sufficiently large that we should consider using a different method for performing exploratory factor analysis, such as the default principle factor method, which does not assume that the uniquenesses are 0. Had we instead typed `factor x1-x9, pf`, the results would have been similar, having only one factor with an eigenvalue greater than 1 and with factor loadings ranging from 0.51 to 0.69 on that factor.

Let us proceed with just the first nine items.

1.4 Alpha reliability for our nine-item scale

The next step is to assess the reliability of our nine-item scale of conservatism. We will use the `alpha` command with three options: the `item` option gives us item analysis, the `label` option includes labels of our variables (which can make output look messy if you have long labels), and the `asis` option (which stands for "as is") does not let Stata reverse-code items to get them to fit better. If you have an item that is coded in the opposite direction, you should reverse-code it yourself before running the analysis.

Here is the `alpha` command with results. Stata can estimate alpha using the variance and covariances (`unstandardized`, the default) or the correlations (`standardized`). Because we are going to generate mean or total scores, we will estimate the unstandardized value. The unstandardized version is recommended when generating a scale score using unstandardized variables.

```
. alpha x1-x9, item label asis
Test scale = mean(unstandardized items)
```

Items	S	it-cor	ir-cor	ii-cov	alpha	label
x1	+	0.664	0.505	.19857	0.789	GOVT RESPONSIBILITY – PROVIDE JOBS 2006
x2	+	0.589	0.454	.21848	0.793	GOVT RESPNSBLTY – KEEP PRICES UND CTRL 2006
x3	+	0.669	0.573	.21577	0.781	GOVT RESPNSBLTY – HLTH CARE FOR SICK 2006
x4	+	0.658	0.568	.21954	0.783	GOVT RESPNSBLTY –PROV ELD LIV STAND 2006
x5	+	0.582	0.441	.21865	0.795	GOVT RESPNSBLTY –PROV IND HELP 2006
x6	+	0.650	0.503	.20456	0.788	GOVT RESPNSBLTY –PROV UNEMP LIV STAND 2006
x7	+	0.656	0.487	.19844	0.793	GOVT RESPNSBLTY –REDUCE INC DIFF 2006
x8	+	0.540	0.441	.2348	0.797	GOVT RESPNSBLTY –PROV COLL FIN AID 2006
x9	+	0.717	0.622	.20509	0.774	GOVT RESPNSBLTY –PROV DECENT HOUSING 2006
Test scale				.21262	0.807	mean(unstandardized items)

Our scale looks great by conventional standards. At the bottom of the table in the row labeled `Test scale`, we have the alpha for our scale. The alpha is 0.81, which is over the 0.70 minimum value standard. Under the column labeled `alpha`, we see what would happen if we dropped any single item from our scale; in each case, the alpha would go down. If dropping an item (one at a time) would substantially raise the alpha, we might look carefully at the item to make sure it was measuring the same concept as the other items. Most likely, the PCFA would have spotted such a problematic item as not fitting the first factor.

To obtain our scale score for each person in our sample, we would simply compute the total or mean score for the nine items. I usually prefer the mean score of the items, because it will be on the same scale as the original items (for these items, between 1 to 4). Given this, a mean of 3.0 would denote that a person is conservative and does not support a proactive government. A mean of 1.5 would denote that the person is fairly liberal, between definitely and probably supporting a proactive government.

By contrast, a total score would range from 9 to 36, and it would be much harder to interpret a total score of, say, 24.0 (instead of 3.0) or 12.0 (instead of 1.5). Another problem with the total score arises if there are missing values for some items. An item with a missing value would contribute nothing to the total, as if we had assigned that item a value of 0.0. If a person skips an item, giving them a score of 0 for that item is ridiculous because that would indicate more definite support of a proactive government than the most favorable available response that is coded as 1.0.

To obtain the mean score for each person, we generate our scale score as the mean of the items the person answered. This **egen** (extended generation) command gives you the mean of however many of the nine items the person answered:

```
. egen conserve = rowmean(x1-x9)
(7097 missing values generated)
```

The **egen** command shows that there are 7,097 missing values on our generated **conserve** variable. This is not a problem because the item was only asked for a subset of the overall dataset. The **summarize** command below tells us that the mean is 1.78, the standard deviation is 0.51, and this is based on 1,888 observations. These 1,888 observations include anybody who answered at least one of the items (see box 2.1 for alternative treatments of missing values). The histogram with a normal distribution overlay (figure 1.1) shows that our score is pretty skewed to the right with a concentration of people favoring a proactive government.

```
. summarize conserve, detail
```

```
                              conserve

            Percentiles     Smallest
   1%            1               1
   5%            1               1
  10%         1.111111           1         Obs                 1888
  25%         1.354167           1         Sum of Wgt.         1888

  50%         1.690476                      Mean            1.775299
                              Largest       Std. Dev.        .5132186
  75%         2.111111       3.888889
  90%         2.444444       3.888889       Variance         .2633934
  95%         2.666667           4          Skewness         .7200074
  99%         3.222222           4          Kurtosis         3.537959
```

```
. histogram conserve, norm freq
(bin=32, start=1, width=.09375)
```

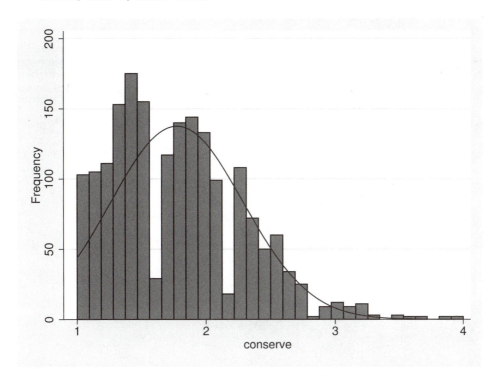

Figure 1.1. Histogram of generated mean score on conservatism

1.5 Generating a factor score rather than a mean or summative score

When we generated our `conserve` scale using the traditional approach, we simply got the mean of the nine items. This method counts each item as equally relevant to the concept being measured. If all the items are equally important, we say that the items are τ ("tau") equivalent. If this were true, then each item would have an identical loading; rarely is this the case. An item that has a loading of 0.90 on a factor is more salient than one that has a loading of 0.30. Therefore, the item with the larger loading should be given a greater weight when we generate the scale score.

You can generate a factor score that weights each item according to how salient it is to the concept being measured. Factor scores will be extremely highly correlated with the simple mean or summative score whenever the loadings are all fairly similar. If the loadings vary widely, the factor score will be a better score to use because factor scores

weight items by their salience (loadings and correlations with the other items), but the
advantage is only substantial when some items have much weaker loadings than others.
The factor score will be scaled to have a mean of 0.0 and a variance of 1.0; in other
words, it will be the standardized score for the concept.

To generate a factor score on the first factor, we run the postestimation command
`predict` immediately after the `factor` command. The `predict` command we use
includes only those participants who answered all nine items. This means we may have
a substantially smaller N if several participants have skipped at least one of the items.
This casewise deletion can be a serious limitation because we normally want to use all
available data.

```
. factor x1-x9, pcf

  (output omitted )

. predict conservf1
(regression scoring assumed)

Scoring coefficients (method = regression)
```

Variable	Factor1
x1	0.16598
x2	0.15641
x3	0.19200
x4	0.18958
x5	0.15468
x6	0.16475
x7	0.16179
x8	0.15866
x9	0.19654

We do not need any options on the `predict` command because the default is
to generate the factor score for the first factor. By contrast, the `egen conserve =
rowmean(x1-x9)` we used in the previous section computed the mean of however many
items a person answered so long as they answered at least one item. We have 1,625
people who answered all nine items and 1,888 people who answered at least one of the
nine items; therefore, the `egen` command retains more observations.[2]

The results above show us the factor scoring coefficients, which are like standardized
beta weights. Notice that the ninth item has a scoring coefficient of 0.20 and the second
item has a scoring coefficient of 0.16. This means the ninth item counts slightly more in
generation of the factor score, which makes sense because the ninth item had a bigger
loading than the second item ($0.74 > 0.59$).

The default for the `predict` command is to predict the factor score as the weighted
sum of the items using the scoring coefficient as the weight for each item. The factor score should be more reliable than the summative or mean score because it more
optimally weights the items.

---

2. To get the mean for only those who answered all nine items, we would have used `egen conservm =
   rowmean(x1-x9) if !missing(x1, x2, x3, x4, x5, x6, x7, x8, x9)`. Stata reads `!missing` as
   "not missing". Notice the items are all listed and separated by a comma; the commas are necessary
   for this command.

How much does it matter whether you compute a mean/summative score or a factor score? The correlation between the average score on conservatism and the factor score is $r = 0.99$. Thus it does not matter which approach you use in this example, except for the different ways of handling missing values on skipped items. Here is a graph comparing the distributions of the two variables. The factor score would be more reliable when the items varied substantially in their loadings and, hence, their factor scoring coefficients. The mean or summative score would use more information if there were a lot of missing values.

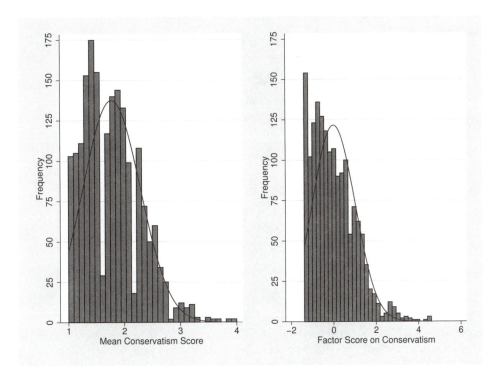

Figure 1.2. Generated mean score on conservatism versus factor score on conservatism

1.6 What can CFA add?

One thing that both CFA and factor analysis methods other than PCFA do is to allow each item to have its own unique variance. This is illustrated in figure 1.3, where each observed question $(x_1$–$x_9)$ has a corresponding error term, ϵ_1–ϵ_9. These error terms allow for variance in the responses to the question that are unique to the item and do not reflecting the shared variance of the nine items. The latent variable, Conservative, appears in the oval and is what the nine items share; the ϵ's are what is unique about each item. It is usual to assume that the error terms are normally distributed and uncorrelated (this assumption will be relaxed in later sections of the book).

CFA assumes that the latent variable accounts for how people respond to all nine individual questions, which is what the nine items share in common. Notice the direction of the arrows from `Conservative` to each of the nine items; the arrows take us from the latent variable to the observed items. This is because how people respond to a question is the dependent variable; that is, a person's response depends on how conservative he or she is, the independent variable. Because all the items seem to tap conservatism, we will posit that a single factor is all we need, and so we draw the single-factor model seen in figure 1.3.

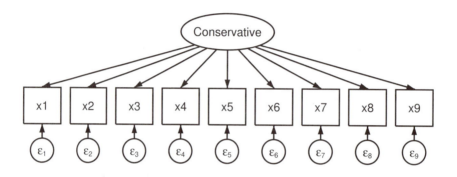

Figure 1.3. CFA for nine-item conservatism scale

This is a confirmatory model because we have specified the factor that underlies the responses to these nine items; that is, all the items are indicators of conservatism. When we ran PCFA, we hoped there would be a single dominant factor. With CFA, we specify the number of factors. In this example, we specified that the covariance of the nine items is fully explained by the single latent variable plus the unique variance of each item. Notice that we are estimating the unique variance or error variance for each of the nine observed indicator variables (items). In the PCFA, we assumed conservatism had to explain all the variance among the nine items. Here we acknowledge that each item may have some unique variance that we are treating as random error. We assume the error variables are normally distributed with a mean of 0.

There are real advantages to CFA. By isolating the shared variance of the nine questions from their unique variances, we are able to obtain a better measure of the latent variable. We are also likely to get stronger results by removing measurement error if the latent variable is subsequently used as an independent or dependent variable in a structural equation model. This is because measurement error, by its nature, only adds noise to our measurement; it has no explanatory power.

Box 1.1. Using the SEM Builder to draw a model

Appendix A provides an introduction to using the extremely capable drawing package that Stata offers, the SEM Builder. Here I will just show how we created figure 1.3. In the Command window, type in `sembuilder` to open the drawing program.

Select the **Add Measurement Component** tool, 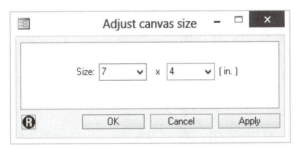, and then click within the SEM Builder wherever you want the latent variable to appear (for our purposes, you will want it to appear in the middle horizontally and a bit high vertically).

In the box labeled *Latent variable name*, type `Conservative` (remember that our convention is to capitalize the first letter of a latent variable). In the box labeled *Measurement variables*, choose the variables `x1` through `x9` from the drop-down menu (assuming you have the `nlsy97cfa.dta` dataset open). Make sure the *Measurement direction* is `Down`. Click **OK**.

With nine indicator variables, the default size for observed variable boxes will cause the diagram to be wider than the default size of the canvas. To see the full diagram, click on the **Adjust Canvas Size** button, , and set the canvas size to 7×4.

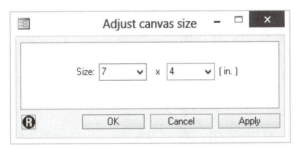

This new canvas size will be large enough to accommodate the full diagram. However, you may not yet be able to see the full canvas. Click on the **Fit in Window** button, , to see the full canvas in the Builder window. If a portion of the diagram is not on the canvas, click on the **Select** tool, , and drag it over the model so that all objects are highlighted. Then move the diagram until you see the entire diagram on the canvas.

Box 1.1. (*continued*)

Because `Conservative` is so long, it does not fit in the default size oval for a latent variable. To make the oval larger, select **Settings** > **Variables** > **All**

Latent... from the menu (on a Mac, click the **Tools** button, [Tools], in the upper right to find the **Settings** menu.) In the dialog box that opens, change the size to 0.75×0.38. You can also change the size of the boxes for observed variables through the **Settings** > **Variables** menu if you like.

Should you need to copy this diagram to another document, such as a Word document, you can do this with the standard copy and paste commands. You can use the **Adjust Canvas Size** button if you want to specify the exact width and height of the object that is copied. Then click on the **Copy Diagram** button,

[icon]. Now the diagram is ready to paste into another document. More detail appears in the chapter 1 appendix.

With so many indicators, it should be clear now why you want short names for your variables. I used the `clonevar` command to rename the variables because their original names in the dataset were long and unclear, for example, `clonevar x1 = s8332500`.

1.7   Fitting a CFA model

We can fit a CFA model by using the Stata command language directly or by using the SEM Builder. Here I will show how to do this with the commands, to ensure that you understand them. The chapter 1 appendix then replicates selected results with the SEM Builder.

The Stata command to fit our CFA model is simple. We do need to run a set of four commands, but each of them is quite simple. First, to fit the model, we run

```
. sem (Conservative -> x1-x9)
```

By running this command, we have the name of our latent variable, `Conservative`, and the `->` points from the latent variable to its indicators, `x1-x9`, just like in figure 1.3.[3] The direction of the arrow is sometimes difficult for beginners to grasp. The idea is that a person's response to each item is caused by how conservative he or she is. That is, your response to an item does not cause you to be conservative; rather, your level of conservatism causes your response. The latent variable here is the independent variable, and the indicators are the dependent variables. We have not specified any options. We have four possible estimation methods:

---

3. Note that the name of the latent variable should be capitalized to help us distinguish indicators, which should be all lowercase, from latent variables.

1. The default is `method(ml)` which means that we fit the model using maximum likelihood estimation. By default, when using `method(ml)`, the variance–covariance matrix of the estimators (and therefore the standard errors) is computed using an observed information matrix. Where you assume normality, `method(ml)` is often the best option and is fairly robust even with some violation of normality. This uses listwise deletion.

2. When option `method(ml)` is combined with option `vce(robust)`, `sem` performs quasi maximum likelihood estimation, and the standard errors are estimated in a manner that does not assume normality. This uses the Huber–White sandwich estimator of the variance–covariance matrix of the estimators. Because several of our items are clearly not normally distributed, this might be a good option to use. The robust standard errors are less efficient than the observed information matrix standard errors if the assumptions of maximum likelihood estimation are met. This uses listwise deletion.

3. The option `method(adf)` is asymptotically distribution free. This method makes no normality assumptions and is a form of weighted least squares. It is also less efficient than maximum likelihood where that is appropriate, but more efficient than the quasi maximum likelihood estimation. Because it does not assume normality and is asymptotically equivalent (in a large sample) to maximum likelihood, this may be the best option for our data. This uses listwise deletion.

4. The option `method(mlmv)` is appropriate when you want to use all the information available in the presence of missing values on one or more variables. This method assumes joint normality and that the missing values are missing at random. This does not use listwise deletion. In our example, we would have an $N = 1888$ using the `method(mlmv)` option, whereas with any of the other three estimators our $N = 1665$.

You can also use the `vce(bootstrap)` option to estimate the standard errors with the bootstrap procedure. This method will resample your observations with replacement and fit the model however many times you specify. It will then use the distribution of the parameter estimates across these replications to estimate your standard error. This will be especially useful when you are concerned about violating the normality assumption of the maximum likelihood options. For example, you might run the following command:

```
. sem (Conservative -> x1-x9), vce(bootstrap, reps(1000) seed(111))
```

We are using the `vce(bootstrap)` option and specifying `reps(1000)`, which means that we are drawing 1,000 samples for our replications. The `seed(111)` option is used so that we can replicate our results; you will get different results each time you run the command unless you set a seed.

For now, we will just use the default version of the command. Here are the results:

```
. sem (Conservative -> x1-x9)
(7360 observations with missing values excluded)
Endogenous variables
Measurement:  x1 x2 x3 x4 x5 x6 x7 x8 x9
Exogenous variables
Latent:       Conservative
Fitting target model:
Iteration 0:   log likelihood = -15604.985
Iteration 1:   log likelihood = -15594.134
Iteration 2:   log likelihood =  -15593.73
Iteration 3:   log likelihood = -15593.729
Structural equation model                    Number of obs      =       1625
Estimation method  = ml
Log likelihood     = -15593.729

 ( 1)  [x1]Conservative = 1
```

	Coef.	OIM Std. Err.	z	P>|z|	[95% Conf. Interval]	
Measurement						
x1 <-						
Conservat~e	1	(constrained)				
_cons	2.329846	.0253521	91.90	0.000	2.280157	2.379535
x2 <-						
Conservat~e	.7377011	.0451423	16.34	0.000	.6492237	.8261784
_cons	1.617231	.0198829	81.34	0.000	1.578261	1.656201
x3 <-						
Conservat~e	.8267157	.0432635	19.11	0.000	.7419209	.9115105
_cons	1.414154	.0167434	84.46	0.000	1.381337	1.44697
x4 <-						
Conservat~e	.7555335	.0403806	18.71	0.000	.676389	.834678
_cons	1.362462	.0155865	87.41	0.000	1.331913	1.39301
x5 <-						
Conservat~e	.7380149	.0462134	15.97	0.000	.6474383	.8285914
_cons	1.769846	.0202603	87.36	0.000	1.730137	1.809556
x6 <-						
Conservat~e	.9146378	.053406	17.13	0.000	.8099639	1.019312
_cons	2.259692	.0229301	98.55	0.000	2.21475	2.304634
x7 <-						
Conservat~e	1.028027	.0614681	16.72	0.000	.9075522	1.148503
_cons	2.219692	.0266439	83.31	0.000	2.167471	2.271913
x8 <-						
Conservat~e	.5486913	.033463	16.40	0.000	.483105	.6142775
_cons	1.307077	.0141374	92.46	0.000	1.279368	1.334786

```
x9 <-
  Conservat~e    .9278118    .0479147    19.36   0.000    .8339008    1.021723
        _cons   1.705231    .0187041    91.17   0.000    1.668571    1.74189

     var(e.x1)    .7287257    .0280851                     .6757076    .7859038
     var(e.x2)    .4706031    .0178489                     .4368885    .5069195
     var(e.x3)    .2397812    .0104761                     .2201029    .2612188
     var(e.x4)    .2145611     .009255                     .1971672    .2334895
     var(e.x5)    .4950753    .0186802                     .4597838    .5330757
     var(e.x6)     .590299    .0229507                     .5469876    .6370399
     var(e.x7)    .8199315    .0314634                     .7605262    .8839769
     var(e.x8)    .2297334    .0087974                      .213122    .2476396
     var(e.x9)    .2967257    .0129788                     .2723476    .3232858
  var(Conserv~e)   .3157048    .0287081                      .264167    .3772973
```

```
LR test of model vs. saturated: chi2(27)  =    419.01, Prob > chi2 = 0.0000
```

1.8   Interpreting and presenting CFA results

At the top of the results, we see that we have 7,360 observations with missing values excluded. The default estimation method, maximum likelihood, uses listwise deletion and drops any observations that do not have a response for all nine of our items.[4] The results next report our endogenous (dependent) variables. All of our observed items, x1 to x9, are endogenous; that is, these measurement variables depend on the latent variable. We next have a list of exogenous variables. Stata reports just one latent exogenous variable, Conservative; Stata does not list the measurement-error terms ϵ_1 to ϵ_9 here even though these are also latent exogenous variables.

The maximum likelihood estimator maximizes the log-likelihood function. Stata converges quite quickly, taking just three iterations. We do not use the log-likelihood function directly. Notice that with listwise deletion, we only have 1,625 observations that have no missing values.

The results above include a section labeled Measurement and a section reporting variances. The measurement section gives estimates of the unstandardized measurement coefficients (factor loadings), their standard errors, and a z test for each estimate along with a 95% confidence interval. By contrast, our PCFA estimates only included factor loadings. The variance section shows the estimates of the variances of the error terms, ϵ_1 to ϵ_9. In the column labeled Coef. appears the unstandardized solution. To identify the variance of the latent variable, Conservative, Stata fixes the loading of the first indicator at 1.0. The indicator that has its loading fixed at 1.0 is called the reference indicator. All the unstandardized estimates will change if you change the reference indicator. It is a good idea to have one of the stronger indicators be the reference indicator. If you want the second indicator to be the reference indicator, you would simply list the variables with that indicator appearing first: sem (Conservative -> x2 x1 x3-x9).

---

4. If we had wanted a full information approach that utilized all available information, we would have specified sem (Conservative -> x1-x9), method(mlmv).

With exploratory factor analysis, we usually focus on the standardized solution. To obtain the standardized solution, we need to replay our results by typing `sem, standardized`. This command simply takes the `sem` output and standardizes it so that all variables, both observed and latent, have a variance of 1.0. Alternatively, we could have listed `standardized` as an option within the original `sem` command, which would yield a standardized solution instead of the unstandardized solution. It is common when reporting your results to report just the standardized values in a figure and both the unstandardized and the standardized values in a table.

```
. sem, standardized
Structural equation model                    Number of obs    =      1625
Estimation method  = ml
Log likelihood     = -15593.729
 ( 1)  [x1]Conservative = 1
```

Standardized	Coef.	OIM Std. Err.	z	P>|z|	[95% Conf. Interval]
Measurement					
x1 <-					
Conservat~e	.549795	.0200518	27.42	0.000	.5104942 .5890958
_cons	2.279751	.0470588	48.44	0.000	2.187518 2.371985
x2 <-					
Conservat~e	.5171478	.0208436	24.81	0.000	.4762951 .5580005
_cons	2.017742	.0432214	46.68	0.000	1.93303 2.102454
x3 <-					
Conservat~e	.6882205	.016421	41.91	0.000	.656036 .720405
_cons	2.09521	.044341	47.25	0.000	2.008303 2.182116
x4 <-					
Conservat~e	.6756463	.0168425	40.12	0.000	.6426355 .7086571
_cons	2.16845	.0454115	47.75	0.000	2.079445 2.257455
x5 <-					
Conservat~e	.5077306	.0210256	24.15	0.000	.4665212 .54894
_cons	2.167021	.0453905	47.74	0.000	2.078057 2.255985
x6 <-					
Conservat~e	.555978	.0200541	27.72	0.000	.5166727 .5952834
_cons	2.444653	.0495404	49.35	0.000	2.347556 2.541751
x7 <-					
Conservat~e	.5378005	.0204071	26.35	0.000	.4978034 .5777977
_cons	2.066659	.0439268	47.05	0.000	1.980564 2.152753
x8 <-					
Conservat~e	.5409708	.0202147	26.76	0.000	.5013507 .580591
_cons	2.29354	.0472646	48.53	0.000	2.200903 2.386177
x9 <-					
Conservat~e	.6914122	.0162749	42.48	0.000	.6595139 .7233104
_cons	2.26162	.046789	48.34	0.000	2.169916 2.353325

var(e.x1)	.6977254	.0220488		.6558217	.7423066
var(e.x2)	.7325582	.0215584		.6914999	.7760543
var(e.x3)	.5263525	.0226025		.4838655	.5725702
var(e.x4)	.5435021	.0227592		.5006764	.589991
var(e.x5)	.7422096	.0213507		.7015209	.7852583
var(e.x6)	.6908884	.0222993		.6485363	.7360063
var(e.x7)	.7107706	.0219499		.6690257	.7551202
var(e.x8)	.7073505	.0218711		.6657569	.7515427
var(e.x9)	.5219492	.0225054		.4796519	.5679763
var(Conserv~e)	1	.		.	.

```
LR test of model vs. saturated: chi2(27)  =    419.01, Prob > chi2 = 0.0000
```

Let us go over these results and figure 1.4 carefully. We now have a loading of each of the nine indicators on our latent variable. Because the variance of the latent variable is fixed at 1.0 for the standardized solution, there is no longer a need for a reference indicator. The standardized loadings are computed by multiplying the unstandardized coefficients by the model-implied standard deviation of the indicator divided by the standard deviation of the latent variable.

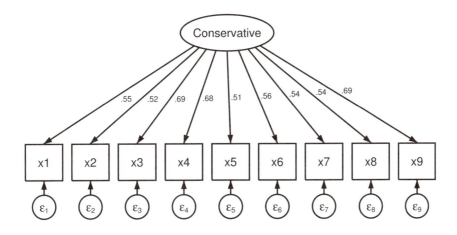

Figure 1.4. Presenting standardized results for a single-factor solution

The standardized coefficient for `Conservative` → `x1` is 0.55 and its standard error is 0.02, so the z statistic (the coefficient/standard error) is 27.42. This is highly significant, $p < 0.001$. The 95% confidence interval for the coefficient is $[0.51, 0.59]$. All the loadings are substantial and statistically significant. Sometimes we place asterisks to indicate the significance of the loadings. In this case, all loadings are significant at the $p < 0.001$ level, so we might simply note this in a footnote. The standardized loadings have a simple interpretation. If you are 1 standard deviation higher on `Conservative`, you will respond 0.55 standard deviations higher on `x1`, 0.52 standard deviations higher on `x2`, etc. They can be interpreted in roughly the same way you would interpret standardized beta weights.

When we are using a CFA model, we can estimate reliability. Reliability refers to the proportion of the total variation in a scale formed by our indicators that is attributed to the true score. If we obtain a high reliability estimate—say, 0.80—this means that our true score can account for 80% of the variation in the scale we are using. The latent variable in the CFA model is intended to represent the true score and be free of any measurement error. We can estimate the reliability of our scale items in a way that accounts for the relative centrality of each item and the error in each item. The alpha coefficient is the lower limit of the true reliability because alpha assumes all indicators have identical centrality equivalent to constraining all the loadings to be equal. The formula for the reliability of the scale measuring our latent variable, ρ ("rho"), is

$$\rho = \frac{\left(\sum \lambda_i\right)^2}{\left(\sum \lambda_i\right)^2 + \sum \theta_{ii} + 2\sum \theta_{ij}}$$

where $\left(\sum \lambda_i\right)^2$ is the squared sum of the unstandardized loadings, called lambdas, and $\sum \theta_{ii}$ is the sum of the unstandardized error variances, called thetas. The $2\sum \theta_{ij}$ is two times the sum of the unstandardized covariances of the errors, if there are any correlated errors in the model. When there are no correlated errors, the equation reduces to

$$\rho = \frac{\left(\sum \lambda_i\right)^2}{\left(\sum \lambda_i\right)^2 + \sum \theta_{ii}}$$

Alpha is the lower limit of ρ when there are no correlated errors, as in our example. Alpha may be greater than ρ when there are correlated error terms, whether these are between indicators of the same latent variable or indicators of different latent variables. You need to use the unstandardized loadings when estimating these reliability values. In addition, to use this formula you need to rerun your structural equation model and fix the variance of the latent variable at 1.0; so far, we have been using a reference indicator to estimate the variance of the latent variable, but we no longer need to do this when we fix the variance of the latent variable at 1.0. In section 1.9.2, we will work out a detailed example of how to do this.

1.9 Assessing goodness of fit

How well does our model fit the data? With CFA, we can get some more information to help us evaluate this. We run a second postestimation command that provides extended statistical information involving the goodness of fit (abbreviated as gof in the command below). We ask for all available statistics by using the stats(all) option.

```
. estat gof, stats(all)
```

Fit statistic	Value	Description
Likelihood ratio		
chi2_ms(27)	419.007	model vs. saturated
p > chi2	0.000	
chi2_bs(36)	3872.316	baseline vs. saturated
p > chi2	0.000	
Population error		
RMSEA	0.095	Root mean squared error of approximation
90% CI, lower bound	0.087	
upper bound	0.103	
pclose	-0.000	Probability RMSEA <= 0.05
Information criteria		
AIC	31241.457	Akaike's information criterion
BIC	31387.075	Bayesian information criterion
Baseline comparison		
CFI	0.898	Comparative fit index
TLI	0.864	Tucker-Lewis index
Size of residuals		
SRMR	0.045	Standardized root mean squared residual
CD	0.835	Coefficient of determination

In the first section of the results, we have a chi-squared of 419.01 with 27 degrees of freedom, $p < 0.001$. This chi-squared value also appeared at the bottom of both our unstandardized and our standardized solutions. The chi-squared compares our model to a saturated model that has no degrees of freedom. The degrees of freedom is based on the amount of information we have and the number of parameters we are estimating. Structural equation modeling is trying to reproduce the covariance matrix for our nine items; it picks the combination of parameter estimates that does the best job of reproducing the covariance matrix.

We can use the sem command to compute covariances. To display the covariance matrix, we can use the postestimation command estat framework. Specifically, we can type sem x1-x9 followed by estat framework, fitted to obtain the covariance matrix for our nine items.

```
. sem x1-x9
```

(output omitted)

```
. estat framework, fitted
(model contains no latent variables)
```

(output omitted)

Fitted covariances of observed variables

Phi	observed x1	x2	x3	x4	x5
observed					
x1	1.044432				
x2	.3016396	.6424108			
x3	.2498545	.2046792	.4555535		
x4	.1992127	.1793545	.2498852	.3947755	
x5	.2482231	.2165196	.1611653	.1806527	.6670292
x6	.3475723	.1387868	.2081397	.19141	.2302307
x7	.37492	.2434761	.2320905	.2028315	.2111784
x8	.1393272	.118155	.1491306	.1465426	.1309826
x9	.2843054	.1899406	.2217721	.1948425	.2170808

Phi	observed x6	x7	x8	x9
observed				
x6	.8544061			
x7	.3817168	1.153581		
x8	.1122545	.1651529	.3247807	
x9	.3362416	.3459893	.1828245	.5684957

Means of exogenous variables

kappa	observed x1	x2	x3	x4	x5
mean	2.329846	1.617231	1.414154	1.362462	1.769846

kappa	observed x6	x7	x8	x9
mean	2.259692	2.219692	1.307077	1.705231

(output omitted)

There are $\{k(k+1)\}/2$ elements in this covariance matrix, where k is the number of items. Thus there are $(9 \times 10)/2 = 45$ elements. It is possible to write an equation for each of these elements. Consider the covariance of the first pair of items, `cov(x1, x2)` = 0.30. Examining figure 1.4 above, we can see that the only connection between x1 and x2 is their common cause, namely, `Conservative`. Thus the predicted covariance based on our model is

$$\text{cov(x1, x2)} = \text{unstandardized loading of x1 on } \texttt{Conservative}$$
$$\times \text{ variance of } \texttt{Conservative}$$
$$\times \text{ unstandardized loading of x2 on } \texttt{Conservative}$$

In this fashion, we can write 45 simultaneous equations. How many parameters are we estimating? For the unstandardized results, we are estimating 8 loadings[5] + 9 error variances + the variance of `Conservative` = 18 parameters. The standardized result is computed from the unstandardized solution. For the standardized result, we have 9 loadings + 9 error variances = 18 parameters. Thus we have $45 - 18 = 27$ degrees of freedom for our chi-squared test.

Without maximum likelihood estimation, we could pick values for the loadings for `x1` and `x2` that perfectly reproduced the covariance of `x1` with `x2`; however, this particular pair of values might not perfectly reproduce the covariance of `x1` or `x2` with `x3`–`x9`. The maximum likelihood solution produces the best combinations of loadings and variances that come as close as possible to reproducing all 45 variances and covariances in the matrix above.

Our $\chi^2(27) = 419.01$, $p < 0.001$. This is highly significant, and this is not an ideal result. We are testing whether our one-factor model perfectly reproduces the entire 45 pieces of information in the covariance matrix—it does not. We fail significantly to reproduce the covariance matrix.

Acknowledging that our model is not perfect, we can examine how closely our one-factor solution comes to fitting the data. The comparative fit index (CFI) is a widely used measure. It compares our model with a baseline model that assumes there is no relationship among our nine observed indicator variables.

$$\text{CFI} = 1 - \frac{\max(\text{our-chi-squared-df}, 0)}{\max(\text{our-chi-squared-df}, \text{baseline chi-squared-df})}$$

In most cases, this reduces to

$$\text{CFI} = 1 - \frac{\text{our-chi-squared-df}}{\text{baseline-chi-squared-df}}$$

where df is the degrees of freedom for the corresponding model.

In other words, the CFI reported by the `estat gof` command above indicates that our model does 89.8% better than a null model in which we assume the items are all unrelated to each other. This is quite an improvement, but depending on the source, the recommended cutoff values should be either 0.90 or 0.95 with the 0.95 cutoff becoming more widely used today.[6] Our one-factor solution is again less than ideal, falling short of either of these criteria.

---

5. Remember: The first loading is fixed at 1.0 as a reference indicator.
6. I do not describe the Tucker–Lewis index, which has the same cutoff values as the more commonly used CFI.

We can also examine the root mean squared error of approximation (RMSEA) as a measure of fit. This measure of fit considers how much error there is for each degree of freedom. The measure penalizes your model for unnecessary added complexity. If you add more paths (loadings, correlated errors, other dimensions), you use additional degrees of freedom. A more complex model may fit better because it capitalizes on chance, while the RMSEA adjusts for this. The RMSEA is defined as

$$\text{RMSEA} = \sqrt{\frac{T/(N-1)}{\text{df}}}$$

where $T = \max(\text{model chi-squared} - \text{df}, 0)$, df is the degrees of freedom, and N is the sample size.

For our data, the RMSEA = 0.10. It is recommended that this be 0.05 for a good fit and less than 0.08 for a reasonably close fit. Thus our model is not a good fit or even a reasonably close fit. Stata also reports a 90% confidence interval for our RMSEA that ranges from a low of 0.087 (still not good) to a high of 0.103 (definitely a bad fit). Though it would be good to report this confidence interval, many researchers do not. The chance of RMSEA being less than or equal to 0.05 (a good fit) if we had population data is $p < 0.001$.

Another index of how well our model fits is the standardized root mean squared residual (SRMR). This is a measure of how close we come to reproducing each correlation, on average. Our results have an SRMR of 0.05. This means that, on average, we come within 0.05 of reproducing each correlation among the nine indicators. The recommended value is less than 0.08, so our 0.05 is a good value. The SRMR can be misleading if you have items that are minimally related; for example, if the average correlation of items is 0.06, then coming within 0.08 would be terrible, but if the average correlation is 0.60, then coming within 0.08 might be considered okay.

Akaike's information criterion and the Bayesian information criterion are not useful for evaluating a single model. They can be used to compare any models that have the same set of variables and are ideal for a special situation where you are comparing two models that have the same set of variables but are not nested in one another. In such cases, the model with the smaller Akaike's information criterion and Bayesian information criterion will be preferred. Finally, the coefficient of determination is an overall summary of how well the model fits. It has a maximum value of 1.0, but there are no standards for how high it should be. A comparison of cutoff values for the various measures of fit is provided by Hu and Bentler (1999).

Box 1.2. How close does our model come to fitting the covariance matrix?

When we ran the `sem x1-x9` command followed by `estat framework, fitted` command we obtained the covariances of the observed variables in our model. We can run one more command, `estat residuals`. This will tell us how close our CFA model comes to reproducing each variance and covariance. Here are the results:

```
. estat residuals

Residuals of observed variables

  (output omitted)

Covariance residuals
```

	x1	x2	x3	x4	x5
x1	0.000				
x2	0.000	0.000			
x3	0.000	0.000	0.000		
x4	0.000	0.000	0.000	0.000	
x5	0.000	0.000	0.000	0.000	0.000
x6	0.000	0.000	0.000	0.000	0.000
x7	0.000	0.000	0.000	0.000	0.000
x8	0.000	0.000	0.000	0.000	0.000
x9	0.000	0.000	0.000	0.000	0.000

	x6	x7	x8	x9
x6	0.000			
x7	0.000	0.000		
x8	0.000	0.000	0.000	
x9	0.000	0.000	0.000	0.000

From the `estat framework, fitted` command, we saw that the `cov(x1, x2)` was 0.30. The residual covariance matrix of 0.07 indicates we were a ways off. On the other hand, the `cov(x1, x3)` = 0.25, and we are only off by −0.01. The smaller the values in the residual covariance matrix, relative to the values in the observed covariance matrix, the better our model fits. A different loading for x1 or x2 might help fit the `cov(x1, x2)` better, but the gain would be offset by fitting other covariances worse. Our solution yields the smallest residual covariance matrix.

1.9.1 Modification indices

So far, the CFA result is telling us that our model is not ideal, but the result does not indicate what we might change about our model. Stata provides postestimation options that can help here, but it is dangerous to rely too much on this information.

The `estat mindices` postestimation command asks for extended statistical information about modification indices. A modification index is an estimate of how much the chi-squared will be reduced if we estimated a particular extra parameter. For example, our model might fit better if we correlated the error terms for a particular pair of variables. Here are the results of this postestimation command:

```
. estat mindices

Modification indices
```

	MI	df	P>MI	EPC	Standard EPC
cov(e.x1,e.x2)	27.046	1	0.00	.0830313	.1417856
cov(e.x1,e.x4)	22.437	1	0.00	−.0549145	−.1388768
cov(e.x1,e.x6)	16.165	1	0.00	.0727568	.1109317
cov(e.x1,e.x7)	8.434	1	0.00	.0615727	.0796558
cov(e.x1,e.x8)	13.662	1	0.00	−.041522	−.1014807
cov(e.x2,e.x5)	16.394	1	0.00	.0526548	.1090874
cov(e.x2,e.x6)	39.084	1	0.00	−.090013	−.1707818
cov(e.x2,e.x9)	11.090	1	0.00	−.0364539	−.0975527
cov(e.x3,e.x4)	147.976	1	0.00	.0889115	.3919898
cov(e.x3,e.x5)	18.664	1	0.00	−.0433453	−.1258053
cov(e.x3,e.x6)	15.379	1	0.00	−.0438055	−.1164353
cov(e.x3,e.x7)	15.291	1	0.00	−.0510741	−.1151874
cov(e.x3,e.x9)	16.559	1	0.00	−.035565	−.133333
cov(e.x4,e.x6)	12.898	1	0.00	−.0375737	−.1055778
cov(e.x4,e.x7)	22.944	1	0.00	−.0586131	−.1397431
cov(e.x4,e.x8)	11.218	1	0.00	.0217217	.0978375
cov(e.x4,e.x9)	30.371	1	0.00	−.0449624	−.1781953
cov(e.x6,e.x7)	29.683	1	0.00	.1041754	.1497409
cov(e.x6,e.x8)	31.435	1	0.00	−.0568032	−.1542497
cov(e.x6,e.x9)	62.385	1	0.00	.0984096	.2351387
cov(e.x7,e.x9)	19.055	1	0.00	.0635887	.128918
cov(e.x8,e.x9)	16.553	1	0.00	.0314136	.1203174

```
EPC = expected parameter change
```

Each parameter we might add uses one degree of freedom. A chi-squared with 1 degree of freedom equal to 3.84 is significant at the 0.05 level. Therefore, for each modification index that is bigger than 3.84, we would significantly improve the fit of the model by reducing chi-squared significantly.[7] There is an enormous risk of capitalizing on chance, and you should only make a change if two conditions are met when you review the modification indices:

---

7. Remember: The model chi-squared is testing whether or not our model fits, so we want to consider changes that reduce the size of chi-squared significantly.

- The modification index is substantial.

- You can theoretically justify adding the parameter to your model.

Also remember that these values are for adding a single parameter. The values are not additive, so if you add a parameter, the modification indices for everything else will change. Best practice is to add only one parameter at a time.

By far, the largest modification index is between x3 and x4.

```
. codebook x3 x4, compact
Variable      Obs Unique      Mean  Min  Max  Label

x3           1874     4   1.416222    1    4   GOVT RESPNSBLTY - HLTH CARE FOR S...
x4           1872     4   1.365385    1    4   GOVT RESPNSBLTY -PROV ELD LIV STA...
```

Both of these items involve a focus on helping people who, because of their circumstance (illness or age), may have limited capacity to make a substantial income. estat mindices indicates that the error terms for these two items should be correlated with an approximate correlation (standardized estimated parameter change) of 0.39. In other words, what is unique about one of these items with respect to Conservative is correlated with what is unique about the other item. x3 and x4 are more correlated with each other than they are with the other seven items (the correlation between this pair of items is 0.59).

We can reexamine the model, allowing for these two items to be correlated. (I do not show the results here.) The model is still a weak fit to the data. This makes us wonder if one or more of the items really should be dropped because they are problematic. Item x2 (the government should be responsible for keeping prices under control) does not really fit what we are trying to measure. A person who is against a proactive government that tries to help people might still support the government having a monetary policy that limits inflation. Item x8 (helping with college funding) might also be something a conservative would see as especially beneficial to his or her own family, even if the family were upper middle class.

We will drop these two variables and keep the covariance between the pair of error terms x3 and x4. Our final model is fit and is evaluated using the following four commands. Notice the inclusion of the covariance() option to provide for the error term of x3 to be correlated with the error term of x4. We name these error terms e.x3 and e.x4. The asterisk tells Stata to allow them to be correlated.

```
. sem (Conservative -> x1 x3-x7 x9), covariance(e.x3*e.x4)
  (output omitted)
. sem, standardized
```

Structural equation model   Number of obs   =   1631
Estimation method = ml
Log likelihood   = -12634.282

 (1)   [x1]Conservative = 1

Standardized	Coef.	OIM Std. Err.	z	P>|z|	[95% Conf. Interval]	
Measurement						
x1 <-						
Conservat~e	.5604752	.0207625	26.99	0.000	.5197814	.601169
_cons	2.280544	.046984	48.54	0.000	2.188457	2.372631
x3 <-						
Conservat~e	.5802223	.0204341	28.39	0.000	.5401722	.6202724
_cons	2.088924	.0441682	47.29	0.000	2.002356	2.175492
x4 <-						
Conservat~e	.5629209	.0209335	26.89	0.000	.5218921	.6039497
_cons	2.168194	.0453242	47.84	0.000	2.079361	2.257028
x5 <-						
Conservat~e	.4955946	.0222097	22.31	0.000	.4520644	.5391249
_cons	2.169155	.0453383	47.84	0.000	2.080294	2.258017
x6 <-						
Conservat~e	.6417291	.0186347	34.44	0.000	.6052057	.6782526
_cons	2.441724	.0494048	49.42	0.000	2.344893	2.538556
x7 <-						
Conservat~e	.5752643	.0202855	28.36	0.000	.5355055	.6150232
_cons	2.066516	.0438438	47.13	0.000	1.980584	2.152449
x9 <-						
Conservat~e	.7268749	.0165553	43.91	0.000	.694427	.7593228
_cons	2.263959	.0467375	48.44	0.000	2.172355	2.355563
var(e.x1)	.6858676	.0232738			.6417357	.7330344
var(e.x3)	.6633421	.0237126			.6184569	.7114848
var(e.x4)	.6831201	.0235678			.6384552	.7309095
var(e.x5)	.754386	.022014			.71245	.7987904
var(e.x6)	.5881837	.0239169			.5431267	.6369785
var(e.x7)	.669071	.023339			.624856	.7164146
var(e.x9)	.4716529	.0240673			.4267639	.5212635
var(Conserv~e)	1	.			.	.
cov(e.x3,e.x4)	.3844886	.0237473	16.19	0.000	.3379447	.4310325

LR test of model vs. saturated: chi2(13) =   56.02, Prob > chi2 = 0.0000

```
. estat gof, stats(all)
```

Fit statistic		Value	Description
Likelihood ratio			
	chi2_ms(13)	56.021	model vs. saturated
	p > chi2	0.000	
	chi2_bs(21)	2870.789	baseline vs. saturated
	p > chi2	0.000	
Population error			
	RMSEA	0.045	Root mean squared error of approximation
90% CI,	lower bound	0.033	
	upper bound	0.057	
	pclose	0.729	Probability RMSEA <= 0.05
Information criteria			
	AIC	25312.564	Akaike's information criterion
	BIC	25431.297	Bayesian information criterion
Baseline comparison			
	CFI	0.985	Comparative fit index
	TLI	0.976	Tucker-Lewis index
Size of residuals			
	SRMR	0.021	Standardized root mean squared residual
	CD	0.792	Coefficient of determination

```
. estat mindices
  (output omitted )
```

Our results are very strong, and the goodness of fit is greatly improved. Our model still has a significant chi-squared of 56.02, which is highly significant with 13 degrees of freedom, $p < 0.001$. Although this means our model is not perfect, the measures of fit are all good: RMSEA = 0.05 and CFI = 0.99.

1.9.2   Final model and estimating scale reliability

We use the formula for scale reliability that includes the covariance of the two error terms in the denominator. We use the unstandardized loadings, and we need to refit our model one last time. We need to have the variance of Conservative fixed at 1.0 so that there is no reference indicator needed; that is, each of the λ_i values is estimated. When we fix the variance of Conservative at 1.0, Stata recognizes that we do not need a reference indicator fixed at 1.0.[8] Here is our command with the additional constraint on the error variance and partial results.

---

8. Whether we use a reference indicator to estimate the variance of the latent variable or fix the latent variable's variance at 1.0, the standardized solution reported in figure 1.5 will be the same.

```
. sem (Conservative->x1 x3-x7 x9), covariance(e.x3*e.x4)
> variance(Conservative@1)
(7354 observations with missing values excluded)
```
  (*output omitted*)

```
Structural equation model                    Number of obs      =        1631
Estimation method = ml
Log likelihood      = -12634.282
 ( 1)  [var(Conservative)]_cons = 1
```

		OIM				
	Coef.	Std. Err.	z	P>|z|	[95% Conf. Interval]	
Measurement						
x1 <-						
Conservat~e	.5731977	.0262634	21.82	0.000	.5217223	.624673
_cons	2.332311	.0253234	92.10	0.000	2.282679	2.381944
x3 <-						
Conservat~e	.3928852	.0173651	22.62	0.000	.3588501	.4269203
_cons	1.41447	.0167666	84.36	0.000	1.381608	1.447332
x4 <-						
Conservat~e	.3541811	.0162626	21.78	0.000	.322307	.3860552
_cons	1.364194	.0155794	87.56	0.000	1.333659	1.394729
x5 <-						
Conservat~e	.4042766	.0213058	18.97	0.000	.3625179	.4460353
_cons	1.769467	.0201988	87.60	0.000	1.729878	1.809055
x6 <-						
Conservat~e	.5942818	.0230962	25.73	0.000	.5490141	.6395494
_cons	2.261189	.0229305	98.61	0.000	2.216247	2.306132
x7 <-						
Conservat~e	.6178507	.0273552	22.59	0.000	.5642355	.6714658
_cons	2.219497	.0265943	83.46	0.000	2.167373	2.271621
x9 <-						
Conservat~e	.5474424	.0182758	29.95	0.000	.5116225	.5832622
_cons	1.705089	.0186488	91.43	0.000	1.668538	1.74164
var(e.x1)	.7173593	.0287599			.6631487	.7760015
var(e.x3)	.3041442	.0124605			.2806769	.3295736
var(e.x4)	.2704286	.0109494			.2497976	.2927635
var(e.x5)	.5019923	.0192822			.4655875	.5412437
var(e.x6)	.5044216	.0218112			.4634339	.5490344
var(e.x7)	.7717982	.0311739			.7130544	.8353815
var(e.x9)	.2675341	.0134682			.2423975	.2952774
var(Conserv~e)	1	(constrained)				
cov(e.x3,e.x4)	.110268	.0091451	12.06	0.000	.092344	.128192

```
LR test of model vs. saturated: chi2(13)  =       56.02, Prob > chi2 = 0.0000
```

The formula for reliability is

$$\rho = \frac{(\sum \lambda_i)^2}{(\sum \lambda_i)^2 + (\sum \theta_{ii}) + 2(\sum \theta_{ij})}$$

For our example,

$$\left(\sum \lambda_i\right)^2 = (0.57 + 0.39 + 0.35 + 0.40 + 0.59 + 0.62 + 0.55)^2 = 12.04$$

$$\sum \theta_{ii} = 0.72 + 0.30 + 0.27 + 0.50 + 0.50 + 0.77 + 0.27 = 3.33$$

$$\sum \theta_{ij} = 0.11$$

Therefore, the reliability is

$$\frac{12.04}{12.04 + 3.33 + 2(0.11)} = 0.77$$

In our final model, shown in figure 1.5, we have included the key measures of goodness of fit along with sample size and reliability.

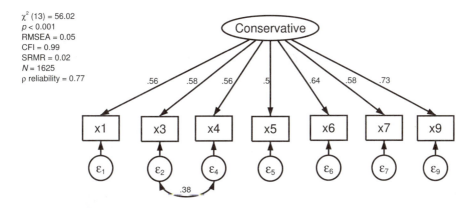

Figure 1.5. Final model

All loadings are significant at the $p < 0.001$ level.

Many readers are perfectly happy to have just a figure showing the results. Other readers want more detailed information, which can be provided in a table. By including a figure as well as a table like table 1.1, you can save space in the text because you will not need to report each value (see box 1.2 for an interpretation of a CFA result).

Table 1.1. Final results for CFA model

	Unstandardized value	Standardized value
Loadings		
x1: Do not provide jobs	1.00 (fixed)	0.56***
x3: Do not provide health care for sick	0.69***	0.58***
x4: Do not provide elderly income	0.62***	0.56***
x5: Do not provide help for indigents	0.71***	0.50***
x6: Do not provide unemployed income	1.04***	0.64***
x7: Do not reduce income differential	1.08***	0.58***
x9: Do not provide decent housing	0.96***	0.73***
Variances		
error.x1	0.72	0.69
error.x3	0.30	0.66
error.x4	0.27	0.68
error.x5	0.50	0.75
error.x6	0.50	0.59
error.x7	0.77	0.67
error.x9	0.27	0.47
Conservative	0.33	1.00 (fixed)
Covariance		
error.x3 with error.x4	0.11***	0.38***

*** $p < 0.001$

This table would be presented along with the measures of fit shown in figure 1.5. The table makes it clear which item we picked as the reference indicator. Notice that Stata does not report significance levels for the error variances, although it does report approximate confidence intervals. Other structural equation model software does report significance levels for the error variances; Stata does not because there is a boundary problem in that a variance cannot be less than 0. However, because none of the confidence intervals for the error variances approaches 0 for its lower limit, we should not dismiss any of these error variances by fixing them at 0. There is a unique variance for each item that is not directly shared with Conservatism.

Box 1.3. Interpretation of CFA results

We used CFA to assess the measurement properties of a scale of conservatism. Our nine initial items all loaded significantly and strongly on a single conservatism dimension, but the fit of the initial model was not satisfactory: $\chi^2(27) = 419.01$, $p < 0.001$, RMSEA $= 0.10$, and CFI $= 0.90$. The significant chi-squared test indicates the model is not a perfect fit. The RMSEA is much higher than its acceptable value of 0.05, and the CFI is lower than its acceptable value of 0.95.

The modification indices indicated several possible changes for our model. Allowing the error terms for a pair of items to be correlated has a substantial modification index value and made conceptual sense. Item x3 concerned health care for the sick, and item x4 concerned the living standard of the elderly. Both of these groups (the sick and the elderly) may have limited capacity to make much income because of either illness or age. By contrast, the other indicators appeared to focus on government action where a conservative might expect the person to be able to take care of his or her own needs. Correlating the x3 and x4 error terms improved our fit, and the correlation was significant; however, the fit remained unsatisfactory.

Further study of the individual items identified a pair of items that were problematic. Item x2 dealt with controlling monetary policy, and item x8 dealt with financial aid for colleges. Both liberals and conservatives may well feel the government should actively influence the monetary policy and may see themselves as beneficiaries of college aid. By dropping these indicators from the models, we obtained a satisfactory fit. The chi-squared with 13 degrees of freedom was still significant ($\chi^2(13) = 56.02$, $p < 0.001$), but the measures of goodness of fit were excellent with RMSEA $= 0.05$ and CFI $= 0.99$.

All seven indicators of conservatism have a substantial loading that is significant at the 0.001 level. The standardized loadings ranged from 0.50 to 0.72. The resulting scale reliability was $\rho = 0.77$.

1.10   A two-factor model

Suppose we have a second dimension, Depress, that is measured by three indicators. In the original NLSY97 dataset, these are measured by t2782800–t2783000; for convenience, we have renamed these x11–x13.

```
. codebook x11-x13, compact
Variable    Obs Unique      Mean  Min  Max  Label
```
Variable	Obs	Unique	Mean	Min	Max	Label
x11	7295	4	3.223715	1	4	HOW OFT R FELT DOWN OR BLUE 2008
x12	7397	4	2.232932	1	4	HOW OFT R BEEN HAPPY PERSON 2008
x13	7291	4	3.655328	1	4	HOW OFT R DEPRESSED LAST MONTH 2008

We have between 7,291 and 7,397 observations. These items were asked of far more people than the items on conservatism. We are interested in evaluating our measure of depression using these three indicators and then combining our latent depression variable and our latent conservative variable to see if they are correlated. We aim to find out whether depressed people are more conservative or less conservative.

1.10.1   Evaluating the depression dimension

A practical way of developing a two-factor model is to first ensure that each factor is meaningful by itself. We have already done this for the Conservative dimension, so let us now do it for the Depress dimension. Is our model for depression identified? We have $3(3 + 1)/2 = 6$ variances and covariances. We have three standardized loadings and three error variances for a total of six parameters. With three indicators, we have a just-identified model. It will be a perfect fit with a chi-squared of 0.0 and 0 degrees of freedom. The CFI will be 1.0 and the RMSEA will be 0.0. Our model is "perfect" because we have no information with which to test its fit. Anytime you have 0 degrees of freedom, your model will be a perfect fit because it is just-identified. In this case, we can fit our model and evaluate the size of the loadings; we just cannot test it. We could not fit a model of depression if we had just two indicators unless we had other latent variables with multiple indicators.

Let us look at the standardized solution to assess our model. We run our set of four commands. But because the estat gof and estat mindices commands do not add any useful information, the output from these two commands is not shown.

```
. sem (Depress -> x11-x13)
```

(output omitted)

```
. sem, standardized
```

```
Structural equation model                      Number of obs    =       7183
Estimation method  = ml
Log likelihood     = -18464.972
 ( 1)   [x11]Depress = 1
```

Standardized	Coef.	OIM Std. Err.	z	P>|z|	[95% Conf. Interval]	
Measurement						
x11 <-						
Depress	.8130901	.0101864	79.82	0.000	.7931251	.8330551
_cons	4.851163	.0421589	115.07	0.000	4.768533	4.933793
x12 <-						
Depress	-.6088417	.0102152	-59.60	0.000	-.6288631	-.5888203
_cons	3.435663	.0309978	110.84	0.000	3.374909	3.496418
x13 <-						
Depress	.654818	.010117	64.72	0.000	.6349891	.6746469
_cons	6.159645	.0527282	116.82	0.000	6.0563	6.26299
var(e.x11)	.3388845	.0165649			.3079245	.3729572
var(e.x12)	.6293117	.0124389			.6053982	.6541699
var(e.x13)	.5712134	.0132496			.5458261	.5977814
var(Depress)	1	.			.	.

```
LR test of model vs. saturated: chi2(0)    =      0.00, Prob > chi2 =      .
. estat gof, stats(all)
```

(output omitted)

```
. estat mindices
```

(output omitted)

The first and third `Depress` indicators are in the correct direction, but item `x12` would need to be reverse-coded if we were using a traditional summated scale. We do not need to do this when constructing a latent variable because the loadings can be either positive or negative for each item. To simplify interpretation, it is always useful to have the majority of the indicators coded so that a higher score is associated with more of the concept. In this case, two of our three items are positively loading on `Depress`, so higher scores for the majority of items reflect greater depression. If this were not the case, we could reverse-code items.

If we examine the standardized solution, we see that all three loadings are significant and their loadings are fairly strong: 0.81, −0.61, and 0.65. The scale reliability of our simple three-item measure of latent `Depress` is $\rho = 0.74$. Remember, to compute ρ, we need to refit the model with the variance of the latent variable fixed at 1.0: `sem (Depress -> x11-x13), var(Depress@1)`. To calculate the reliability estimate, we use the absolute value of the unstandardized loadings.

Box 1.4. Working with missing values

In our measurement model for `Conservative`, only a subsample of people were asked the set of items. Everybody was asked the three items used for measuring `Depress`. When we fit our two-factor model, we might like to include people who answered at least one of the items for `Conservative` and at least one of the items for `Depress`. However, we may want to exclude people who were not asked any of the items related to `Conservative`. We need to have this apply for our analysis of `Depress` by itself, because we are going to have this apply when we put `Depress` and `Conservative` together. Let us review our options and what they do.

Option 1. `sem (Depress -> x11-x13)`

Option 1 uses the default, which is listwise deletion. This will include the 7,183 people who answered all three of the `Depress` indicators; it will exclude people who answered just one or two of the `Depress` indicators. More importantly, it will include thousands of people who were not asked any of the indicators of `Conservative`.

Option 2. `sem (Depress -> x11-x13), method(mlmv) allmiss`

Option 2 uses the `method(mlmv)` estimator. This will include the 7,429 people who answered at least one of the `Depress` items, regardless of whether they had missing values coded as dots or as special missing-value codes (`.a`, `.b`, `.c`, etc.). This will handle part of our problem in that it will include data for those who answered any of the three `Depress` items; the problem is that it will also include people who were not asked any of the `Conservative` items (remember, those items were only asked of a subsample of survey participants).

The `allmiss` option is included here to illustrate a special situation. All the variables `x11`–`x13` have missing values coded as dots. In other cases, however, missing values may have different coding based on why the value is missing. For example, `.a` might be used if a value is missing because the person refused to answer; `.b` might be used because a person said they did not know how to respond; `.c` might be included to indicate an interviewer error. The `allmiss` option includes all these different types of missing values. Simply including `method(mlmv)` will only work for those who just have a dot for missing values; it will throw out anybody who has a special code. If you include the `allmiss` option, those with special codes for missing values will be included. Use caution with the `allmiss` option because a special code might be used if a question was not applicable, and you probably do not want to include those.

Box 1.4. (*continued*)

Option 3. `sem (Depress -> x11-x13) if !missing(x1, x2, x3, x4, x5, x6, x7, x8, x9)`

Option 3 does listwise deletion for the three indicators of `Depress` (the default), but also does listwise deletion for anyone missing one or more of the `x1-x9` indicators used to measure `Conservative` (the `!missing` means "not missing"). This results in a sample size of 1,466 who answered all nine indicators of `Conservative` and all three indicators of `Depress`.

Option 4. `sem (Depress -> x11-x13) if x1 != . | x2 != . | x3 !=. | x4 !=. | x5 !=. | x6 !=. | x7 !=. | x8 !=. | x9 !=., method(mlmv) allmiss`

Option 4 uses full-information maximum likelihood estimation for people who answered at least one of the three indicators of `Depress` and at least one of the nine indicators of `Conservative`. It results in a sample size of 1,752. This strategy uses all available information without including people who were excluded by the researchers' sample design from any of the items used to measure `Depress`.

If we wanted to use `method(mlmv)` and there were not sample design restrictions related to who was asked certain sets of questions, then the strategy for dealing with missing values in option 2 may be the best solution. However, we would need to be comfortable with the assumption of joint normality of our indicators as discussed in section 1.7.

1.10.2 Estimating a two-factor model

We have established a reasonable measurement model for each of our latent variables, `Depress` and `Conservative`. Now we need to solve our measurement model simultaneously for both sets of items. Let us look at the first two of our standard set of four commands, with the estimating command combining two separate CFAs:

```
. sem (Depress -> x11-x13)          /// Measurement of Depress
   (Conservative -> x1 x3-x7 x9),   /// Measurement of Conservative
   covariance(e.x3*e.x4)            //  Covariance of epsilon-3 with epsilon-4

   (output omitted)

. sem, standardized                 //  Provide a standardized solution
Structural equation model                      Number of obs     =      1466
Estimation method  = ml
Log likelihood     =  -15106.46
 ( 1)  [x11]Depress = 1
 ( 2)  [x1]Conservative = 1
```

Standardized	Coef.	OIM Std. Err.	z	P>|z|	[95% Conf. Interval]	
Measurement						
x11 <-						
Depress	.8110982	.0224636	36.11	0.000	.7670703	.8551261
_cons	4.837057	.0930701	51.97	0.000	4.654643	5.019471
x12 <-						
Depress	-.6145463	.0223827	-27.46	0.000	-.6584156	-.570677
_cons	3.429731	.0685134	50.06	0.000	3.295447	3.564015
x13 <-						
Depress	.6472538	.0226055	28.63	0.000	.6029477	.6915598
_cons	6.320635	.1196151	52.84	0.000	6.086193	6.555076
x1 <-						
Conservat~e	.5548683	.0219522	25.28	0.000	.5118427	.5978939
_cons	2.295094	.0497862	46.10	0.000	2.197515	2.392673
x3 <-						
Conservat~e	.5784576	.0214921	26.91	0.000	.5363339	.6205812
_cons	2.08811	.0465751	44.83	0.000	1.996824	2.179395
x4 <-						
Conservat~e	.5652226	.0219167	25.79	0.000	.5222666	.6081786
_cons	2.159795	.047677	45.30	0.000	2.066349	2.25324
x5 <-						
Conservat~e	.4823727	.0236607	20.39	0.000	.4359985	.5287469
_cons	2.185023	.0480674	45.46	0.000	2.090812	2.279233
x6 <-						
Conservat~e	.6633144	.0189403	35.02	0.000	.6261921	.7004367
_cons	2.449348	.0522329	46.89	0.000	2.346973	2.551722
x7 <-						
Conservat~e	.5807723	.0211474	27.46	0.000	.5393241	.6222206
_cons	2.068596	.0462771	44.70	0.000	1.977894	2.159297
x9 <-						
Conservat~e	.730317	.0172026	42.45	0.000	.6966006	.7640335
_cons	2.274796	.0494675	45.99	0.000	2.177842	2.371751

var(e.x11)	.3421197	.0364404			.2776602	.4215437
var(e.x12)	.6223329	.0275104			.5706833	.678657
var(e.x13)	.5810625	.029263			.5264478	.6413432
var(e.x1)	.6921212	.0243612			.6459838	.7415538
var(e.x3)	.6653869	.0248645			.6183952	.7159494
var(e.x4)	.6805234	.0247757			.633656	.7308573
var(e.x5)	.7673166	.0228266			.7238566	.8133859
var(e.x6)	.560014	.0251267			.5128698	.6114918
var(e.x7)	.6627035	.0245637			.6162667	.7126394
var(e.x9)	.4666371	.0251267			.4198993	.5185771
var(Depress)	1	.			.	.
var(Conserv~e)	1	.			.	.
cov(e.x3,e.x4)	.3797951	.0250967	15.13	0.000	.3306065	.4289837
cov(Depress, Conservative)	.115046	.0334032	3.44	0.001	.0495768	.1805152

LR test of model vs. saturated: chi2(33) = 115.44, Prob > chi2 = 0.0000

The first three rows provide the measurement model for Depress. We see standardized loadings of 0.81 for Depress → x11, −0.61 for Depress → X12, and 0.65 for Depress → x13, which are all highly significant. These are slightly different from the values we obtained when we analyzed Depress without the Conservative items. Stata is trying to pick values for each loading that allow us to reproduce the entire covariance matrix, not just the covariances among indicators of each latent construct. The solution we obtained when we just analyzed Depress did not account for the relationship between these three items and the seven indicators we have for Conservative. We also used different observations when fitting the Depress model by itself.

The standardized loadings for the measurement of Conservative appear next. These vary from a low of 0.48 for Conservative → x5 to a standardized loading of 0.73 for Conservative → x9. All these loadings are significant.

Because our solution is standardized, all variables including the latent variables have been rescaled to have a variance of 1.0. This means that the covariance and the correlation between any pair of variables will be the same value. We can think of a correlation as a standardized covariance. The results indicate that the covariance between Depress and Conservative is 0.12, $z = 3.44$, $p < 0.001$. This value is actually the correlation between our two latent constructs, $r_{\text{Depress,Conservative}} = 0.12$. There is a small but statistically significant correlation between how conservative you are and how depressed you are. This correlation tells us nothing about the possible direction of influence between these constructs: we do not know whether being more conservative makes you more depressed or whether being more depressed makes you more conservative. Indeed, this relationship could be entirely spurious if there were some third variable that caused both depression and conservatism.

How well does our model fit the data? Let us find out with our third command:

```
. estat gof, stats(all)
```

Fit statistic	Value	Description
Likelihood ratio		
chi2_ms(33)	115.438	model vs. saturated
p > chi2	0.000	
chi2_bs(45)	3630.536	baseline vs. saturated
p > chi2	0.000	
Population error		
RMSEA	0.041	Root mean squared error of approximation
90% CI, lower bound	0.033	
upper bound	0.050	
pclose	0.958	Probability RMSEA <= 0.05
Information criteria		
AIC	30276.919	Akaike's information criterion
BIC	30446.209	Bayesian information criterion
Baseline comparison		
CFI	0.977	Comparative fit index
TLI	0.969	Tucker-Lewis index
Size of residuals		
SRMR	0.030	Standardized root mean squared residual
CD	0.952	Coefficient of determination

The $\chi^2(33) = 115.44$, $p < 0.001$, says our model fails significantly to account for the information in the observed covariance matrix. However, RMSEA = 0.04, CFI = 0.98, and SRMR = 0.03 are all very good. This model provides a reasonably good fit to the data.

Should we modify our model? The modification indices from our fourth command provide clues:

```
. estat mindices
Modification indices
```

	MI	df	P>MI	EPC	Standard EPC
Measurement					
x12 <-					
Conservative	6.017	1	0.01	.0728604	.0636661
x13 <-					
Conservative	16.582	1	0.00	.1068147	.1045476
x6 <-					
Depress	5.472	1	0.02	-.1008918	-.0591364
x7 <-					
Depress	4.911	1	0.03	.1165089	.0583792
cov(e.x11,e.x12)	16.582	1	0.00	-.2994507	-1.496611
cov(e.x11,e.x13)	6.017	1	0.01	-.18059	-1.046269
cov(e.x11,e.x3)	6.148	1	0.01	.0179504	.0830443
cov(e.x11,e.x6)	4.092	1	0.04	-.0211057	-.0783812
cov(e.x13,e.x4)	5.704	1	0.02	.0149765	.0646806
cov(e.x1,e.x3)	5.651	1	0.02	.0305168	.064618
cov(e.x1,e.x9)	18.483	1	0.00	-.0690142	-.1568485
cov(e.x3,e.x6)	4.246	1	0.04	-.0231679	-.0605392
cov(e.x4,e.x5)	26.360	1	0.00	.0503241	.1347719
cov(e.x5,e.x7)	7.188	1	0.01	-.0502338	-.0801054
cov(e.x6,e.x9)	6.477	1	0.01	.0383348	.1075154

```
EPC = expected parameter change
```

The first set of modification indices is for the measurement loadings. The biggest modification index among this set is 16.58 for Conservative → x13. If we added this path, our model would be more complex conceptually because this item would be loading on two latent variables. There are cases where a single item should load on two latent variables, but this makes the latent variables at least somewhat factorially confounded.

What should we do about x13? The item asks "How often R depressed last month"? This has face validity as an indicator of Depress. I do not see any face validity for it as an indicator of Conservative, and so would not feel justified in allowing the item to load on both dimensions. If you wanted to let x13 load on both dimensions, your command would be sem (Depress -> x11-x13) (Conservative -> x1 x3-x7 x9 x13); you simply have the item loading on both latent variables.

At the bottom of the results, we see that there are several error terms we could correlate. Although all these modification indices are greater than 3.84, we should not make any changes unless we have a compelling reason to do so. Given how well our model fits the data and that logically we do not see why these errors should be correlated, we will go with this model. Figure 1.6 shows our standardized results. All reported values are significant at the $p < 0.001$ level except the correlation of Depress and Conservative, which is significant at the $p < 0.01$ level.

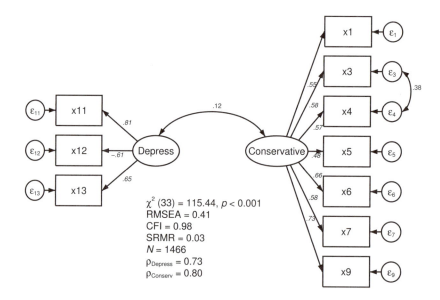

Figure 1.6. Two-factor CFA, final model

The loadings of all indicators of both latent variables are considered strong, that is, greater than 0.40. Notice how the loadings of the indicators of Conservative changed slightly when we combined our items from the full model. This is expected. Structural equation modeling uses what some have called a full-information maximum likelihood estimate in the sense that you pick parameter estimates that optimize the fit between the actual covariances and the estimated covariances based on your full model. We have added several covariances along with three additional variances. With just the seven indicators, we had $(7 \times 8)/2 = 28$ variances and covariances. By adding the three indicators of depression, we now have $(10 \times 11)/2 = 55$ variances and covariances. The optimal parameter estimates will change as we need to account for these additional variances and covariances. Fortunately, they do not change much.

So who is better adjusted: conservatives or liberals? The correlation between Depress and Conservative is 0.12, $p < 0.01$. Thus conservatives are more depressed than liberals, but not much more. The correlation is statistically significant but substantively very weak.

If you have experience with exploratory factor analysis using Stata, you will remember that there are oblique solutions that allow the factors to be correlated. What we are doing with CFA is a bit different. We are estimating the correlation directly in the model, whereas conventional oblique solutions make arbitrary assumptions to get a moderate correlation. If you change those arbitrary assumptions, you can make the correlation either bigger or smaller, but it is questionable to interpret the correlation as very meaningful given it is based on arbitrary assumptions. By contrast, with CFA we can directly estimate the correlation. If our model is correct, the correlation between depression and conservatism is statistically significant but weak at $r = 0.12$.

Box 1.5. What you need to know about identification

Keep in mind that this box is about what you need to know and not what is nice to know. Identification can be simple and often is for many models, but it is sometimes very difficult. One solution is to follow some general rules and then assume that Stata would not give you a definite answer if there were an identification problem. Ideally, you would generate a system of simultaneous equations—one for each variance and covariance—and then determine algebraically whether you have enough information to identify each parameter you are estimating.

What is the issue? You need to have enough pieces of information to identify your parameters. The information available are the variances and covariances of the observed variables. There are $\{k \times (k+1)\}/2$ pieces of information if we have k observed variables in the model. When we analyzed Depress using three indicators by themselves, we had exactly 0 degrees of freedom. That is, we had as many pieces of information (variances and covariances) as we had parameters to be estimated: $(3 \times 4)/2 = 6$ pieces of information, which is the same as the number of estimated variances, covariances, and loadings. We were okay in that we could fit the model with 0 degrees of freedom, but we were unable to test the fit of the model using chi-squared or any of the measures of goodness of fit. When you have 0 degrees of freedom, you have a just-identified model, which must fit the observed covariances perfectly. This is the same issue you had in elementary algebra. You needed one equation to just-identify one unknown, you needed two equations to just-identify two unknowns, and so on. The first rule you should remember is that you need at least as many pieces of information as you have parameters you want to estimate.

Box 1.5. (*continued*)

The sem results also report an intercept for each indicator. We have not used these intercepts yet, but they are identified by the observed means rather than the observed covariances.

What happens if you have more than three indicators for a single latent variable? With four indicators, we would have $(4 \times 5)/2 = 10$ variances and covariances. How many parameters are we trying to estimate? These are easiest to count for the standardized solution. We would need to estimate four loadings (paths from the latent variable to the indicators) plus four error variances. Thus we have 10 pieces of information and 8 parameters to estimate, resulting in $10 - 8 = 2$ degrees of freedom. What if we correlated a pair of error terms? We would still have 10 pieces of information, but now we would have 9 parameters to estimate, meaning that we would have just 1 degree of freedom. We could fit this model and test the fit, although 1 degree of freedom is a fairly limited amount of information for the testing of the model.

What if we had just two indicators of a latent variable, say, a patient's report of her own health and her physician's report? We would have $(2 \times 3)/2 = 3$ variances and covariances. We would need to estimate two loadings and two error variances for a total of four parameters. We cannot estimate four parameters with three pieces of information.

Perhaps you have read articles that use structural equation modeling with several latent variables, one or more of which had just two indicators. In terms of identification, we can estimate a latent variable that has just two indicators if we have additional observed variables in the model.

The counting rule of having more pieces of information than parameters to be estimated does not always work. When solving just-identified simultaneous equations in algebra, we had to have equations that were independent. This same idea can make it difficult to know whether some models are identified in structural equation modeling. This usually happens when there is some sort of feedback. This occurs in structural equation models that are called nonrecursive models, which are covered in the next chapter.

What is better about structural equation modeling than our elementary algebra exercises is that when we have more equations (pieces of information) than unknowns, we can still get a solution. Instead of a just-identified algebraic solution, we get the maximum likelihood solution for an overidentified model. We then use the discrepancy between the observed covariance matrix and the model-implied covariance matrix as our standard for evaluating the goodness of fit.

1.11   Parceling

Sometimes we have so many indicators that the model becomes difficult to fit and impossible to draw. We might have three latent variables with 20 items for each of them. This means we need to estimate $60 - 3 = 57$ loadings, 60 error variances, and the variances and covariances of the latent variables themselves. In cases like this, parceling is one solution you might consider. This approach combines the items into a few parcels. For example, we might have three parcels for each latent variable with the first and second parcels based on seven items and the third parcel based on six items.

How do we combine items into parcels? If we randomly assigned the items to three sets and generated a mean score for each set, we would have three parcels. We could create the parcels like this: `egen parcel1 = rowmean(x1 x2 x3 x4 x5 x6 x7)`. If there were a single underlying dimension for each of the latent variables, then the three randomly assigned sets of items would be approximately equivalent.

A second approach is to actively balance the items going into each parcel. First, you run a separate factor analysis for each of the three latent variables. You then assign the top three items (based on loadings) to anchor each parcel. The next step is to assign the next best triad of items to the three parcels, and continue until you run out of items. It is okay if one of the parcels has fewer items than the others. Balancing the items may be better than random assignment because it ensures that each parcel is a well-balanced representation of the latent variable. As with the random selection, this approach assumes there is a single underlying dimension.

Parceling has several advantages (see Little et al. [2002]). Individual sources of error variance that are not your primary focus are canceled out. Sets of several items going into a parcel will balance out these specific error sources, which can simplify the estimation and yield a better fit. Solutions where you include a large number of individual items are more likely to have convergence problems and to have results that would be unstable if tried on a different sample.

Another time to consider parceling is when you have several highly skewed items but the skew is in opposite directions. By including items with the opposite skew in the same parcel, you may be able to have a more normally distributed observed variable.

1.12   Extensions and what is next

Stata's reference manual on structural equation modeling has numerous examples, and you can apply these to extend what we have done in this first chapter. For example, we could have more latent variables. When we are evaluating the construct validity of a concept, we will include two or more positively related concepts and additional negatively related concepts. We hypothesize that our concept will have a statistically significant positive correlation with concepts that should be related positively and a significant negative correlation with concepts that should be related negatively.

In our two-factor solution, it makes sense to have each indicator paired with just one of the latent variables. In some cases, it may be useful to allow selected indicators to be paired with more than one latent variable. For example, you might be measuring both physical and psychological well-being. You might have several items that relate to physical well-being and several different items that relate to psychological well-being. However, it is possible to have an item that is related to both types of well-being and should load on both latent variables. An item asking if you have severe arthritis that makes you want to give up on life would be an example of an indicator of both dimensions.

We could have error terms for indicators of one latent variable be correlated with error terms for another latent variable. For example, if we measured marital happiness for a set of wives and their husbands, the unique variances for some indicators of the wife's happiness might be correlated with unique variances for some indicators of the husband's happiness.

We could have a methods factor where similar methods are used for some of the items for different latent variables. If we were interested in a latent variable to represent a person's health, we might have five self-report items, a performance measure of fitness, and a clinical assessment. Suppose our second latent variable was compliance and it had a similar mixture of indicators. In addition to our two latent variables, `Health` and `Compliance`, we might have to add a factor for self-report method. This factor would load on the five items that are self-reported health and the five items that are self-reported compliance.

The big extension of CFA is when we use latent variables in full structural equation models, where we replace correlations of latent variables with causal paths between them. Before we look into that, we must first examine using structural equation modeling for path models that do not involve latent variables. Then, chapter 3 will bring these two traditions, CFA and path analysis, together to show how we fit full structural equation models.

1.13 Exercises

1. Why should a measure represent a single dimension?

2. Why is alpha reliability sometimes described as the lower limit on reliability?

3. Why is an average of a series of items easier to interpret than the summation of those items?

4. Use the SEM Builder to draw a three-factor CFA with three indicators of each factor.

5. What is the difference between `method(ml)`, `method(ml)` with `vce(robust)`, and `method(mlmv)`?

6. When would you use `bootstrap` to estimate standard errors?

7. The reliability ρ of a scale corresponding to a latent variable is usually bigger than alpha reliability for the set of indicators. When could ρ be smaller and why?

8. You have 15 observed indicators. What is the maximum number of parameters you can estimate?

9. If you have two results that have modification indices of 40.0, what does this mean and what should you do?

10. You have a psychological measure of competitiveness consisting of 40 items and a measure of liberalism consisting of 30 items. If you are doing a CFA of these two latent variables, what is the problem? What is a solution?

11. You read an article that does a CFA involving three latent variables with three indicators of each. There is an appendix that shows a correlation matrix and standard deviations:

	$x1$	$x2$	$x3$	$x4$	$x5$	$x6$	$x7$	$x8$	$x9$
$x1$	1.0								
$x2$	0.7	1.0							
$x3$	0.6	0.6	1.0						
$x4$	0.2	0.1	0.2	1.0					
$x5$	0.1	0.2	0.1	0.6	1.0				
$x6$	0.1	0.1	0.1	0.6	0.7	1.0			
$x7$	0.3	0.3	0.3	0.1	0.1	0.1	1.0		
$x8$	0.3	0.3	0.3	0.2	0.1	0.1	0.8	1.0	
$x9$	0.3	0.3	0.3	0.1	0.1	0.1	0.9	0.8	1.0
sd	5.4	4.3	3.2	4.1	4.4	5.1	4.4	3.3	4.9

Read appendix B at the end of the book to learn how to enter these data.

    a. Estimate the CFA model using $N = 100$. Carefully interpret the result. Does your model fit the data? How well does it fit? Write out the command rather than using the SEM Builder.

    b. Are you confident these are three latent variables? How correlated are they?

    c. Estimate the model again using $N = 600$—same data, different sample size. Carefully interpret the result. Does your model fit the data? How well does it fit?

    d. Explain the difference between the results for part a and part c.

12. Read the appendix to this chapter. Use the SEM Builder to draw and fit the model for exercise 11.

    a. Show the figure you draw before fitting it.

    b. Show the figure with all the results Stata displays.

    c. Add information about the model fit and delete results that you do not want.

13. You are interested in the relationship between confidence in the physician and patient compliance. You have two measures of confidence: a) a scale of the patient's rating of the physician's medical expertise and b) a scale of the patient's rating of how caring the physician is. You have two measures of patient compliance: a) compliance about prescriptions and b) compliance to behavioral instructions given by the physician. You have measured all four variables at two time points. Your (hypothetical) data are in the `compliance.dta` dataset. Your model looks like this:

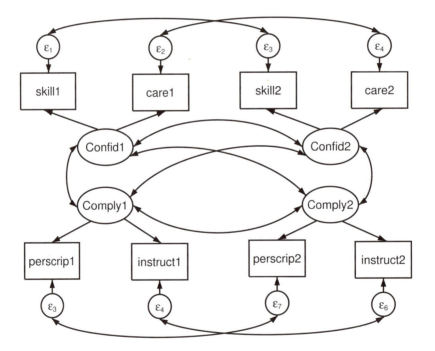

    a. Why are some of the errors correlated? Justify these correlations.

    b. Estimate this model by using the `sem` command in a do-file.

    c. Using the SEM Builder, draw the model and fit it.

1.A   Using the SEM Builder to run a CFA

1.A.1   Drawing the model

We will draw a modified version of the two-factor model shown in figure 1.6 and fit the results using the SEM Builder. We have seven indicators of `Conservative`, which we arranged vertically in figure 1.6. With a large number of indicators, a vertical arrangement usually works best. Here we will arrange the indicators horizontally to illustrate how to handle a large number of indicators horizontally; this is sometimes necessary when there are several latent variables. First, let us open the dataset:

```
. use http://www.stata-press.com/data/dsemusr/nlsy97cfa.dta, clear
```

We open the SEM Builder by typing `sembuilder` in the Command window. (Expand the screen size if you like; you can also click the **Fit in Window** button, , to expand the canvas to the size of the window. Because the names of our indicators are short, we can reduce the default width of the boxes we will use for them. Go to **Settings** > **Variables** > **All Observed...**, change the size to .4 x .38, and click on **OK**. Our `Conservative` latent variables has such a long name, we will also need to change the width of the ovals we will use for latent variables. This time, go to **Settings** > **Variables** > **All Latent...**, change the size to 1 x .38, and click on **OK**.

To create the measurement model for `Depress`, click on the **Add Measurement Component** tool, , on the left side of the screen. Click where you want the oval for the latent variable: you might pick a spot near the center horizontally and above the center vertically. A dialog will open (see figure 1.7).

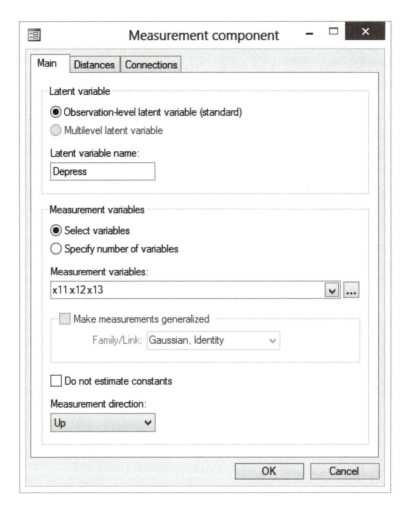

Figure 1.7. Measurement component dialog box

In the *Measurement component* dialog box, you should enter the name of the latent variable, click on *Select variables*, choose the names of the appropriate indicators from the *Measurement variables* control, and change the *Measurement direction* to Up. After you click **OK**, the measurement model appears in your SEM Builder. If you want to move it, click on the selection arrow at the top left of your screen and make sure the full measurement model is selected. Then you can move the measurement model around by clicking and dragging any shape, such as the oval.

Repeat this process for `Conservative` except change the *Measurement direction* to

`Down`. Finally, use the **Add Covariance** tool, 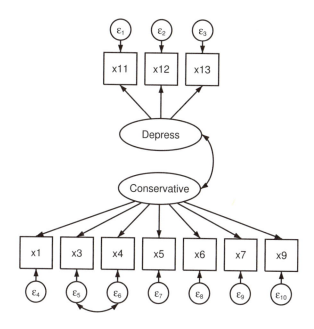, to add curved lines between `Depress` and `Conservative` and between ϵ_5 and ϵ_6. Stata assigns numbers to the error terms based on the order in which you entered the variables. We set up `Depress` first with `x11`, `x12`, and `x13`; thus these items were assigned ϵ_1–ϵ_3. When we entered `x1`, `x3`, `x4`, `x5`, `x6`, `x7`, and `x9` for `Conservative`, ϵ_4–ϵ_{10} were assigned to them. Your model should look similar to figure 1.8. You can see that correlating the errors associated with `x3` and `x4` means we are correlating ϵ_5 with ϵ_6.

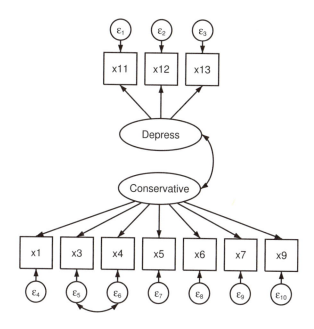

Figure 1.8. Two-factor conceptual model

Box 1.6. Differences between Windows and Mac

Stata does a remarkable job of making Windows, Mac, and Unix interfaces the
same. However, at the time of this writing, there is one interface difference that
is trivial once you are aware of it, but may make some directions a bit confusing.
The top bar for Windows and Unix looks like this:

The top bar for Mac looks like this:

On the Windows and Unix interfaces, you can access the menus at the top: **Object**,
Estimation, **Settings**, and **View**. On the Mac interface, you see a **Tools** button,

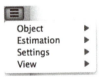, at the top right of the screen. If you do not see the button, you need to right-
click on the bar, select **Customize**, and follow the instructions to get the button
to show up. When you click on the **Tools** button, a drop-down menu appears that
allows you to access the **Objects**, **Estimation**, **Settings**, and **View** menus that
are directly accessible on the top bar of Stata for Windows and Unix.

Both styles of top bar also have other tools available. You can use **Zoom** to adjust
the size of the model you are creating, which is helpful if part of your model has,
say, several arrows in close proximity. The **Fit in Window** tool is also useful.
You can drag the border of the SEM Builder window to be as large or small as you
want, and then you can click on **Fit in Window** to have the canvas fill the entire
window size.

1.A.2   Estimating the model

Now we need to fit the model. Click on **Estimation** > **Estimate...** and then simply click on **OK** to get the results shown in figure 1.9.

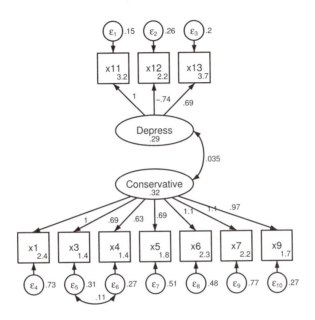

Figure 1.9. Unstandardized solution

This is not a bad model, but it shows several numbers we do not need, such as the variance of Depress and Conservative as well as the intercepts for each observed variable. To remove these, click on **Settings** > **Variables** > **All Latent...**. Select the **Results** tab (see figure 1.10). Both of our latent variables are exogenous, so change the first result for *Exogenous variables* from Variance to None.

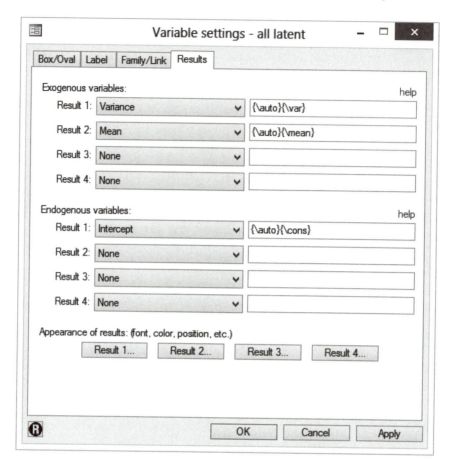

Figure 1.10. Variable settings dialog box

Similarly, click on **Settings** > **Variables** > **Observed Linear Endogenous...** and select the **Results** tab. Stata wants to report an intercept for each observed endogenous variable, but we do not wish to see these intercepts; change this to None. I do not show the results here.

Usually, we want a figure that has the standardized solution. We obtain this by clicking on **View > Standardized Estimates**. Figure 1.11 presents our final drawing.

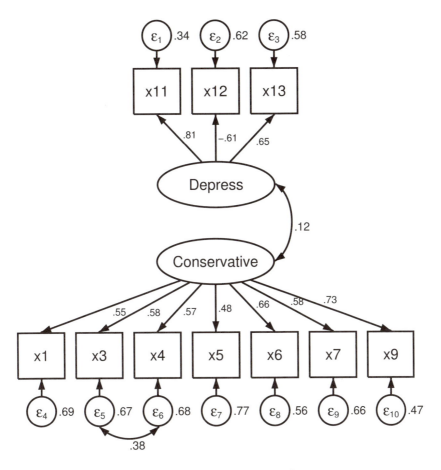

Figure 1.11. Presentation figure

There is one more thing we might do to make the figure a bit more useful for our readers. We can add a legend to report the goodness-of-fit information. Let us put this legend in a box in the left middle of the figure where there is some white space. Click on the **Add Text** tool, **T**, on the left side of your screen and then click on the figure where you want the text to start. Creating the legend can be a bit tricky because we want to italicize the p and the N. We also want the Greek letter ρ (rho). We need to enter {it: p} for the p, {it: {&rho}} for the ρ, and {it: N} for the N. Notice that we are using curly braces rather than parentheses or square brackets.

Here is what we would type in:

```
Chi-square(33) = 115.44
{it: p} < 0.001
RMSEA = 0.41
CFI = 0.98
SRMR = 0.03
{it: {&rho}} (Depress) = 0.73
{it: {&rho}} (Conservative) = 0.80
{it: N} = 1466
```

Our results appear in figure 1.12

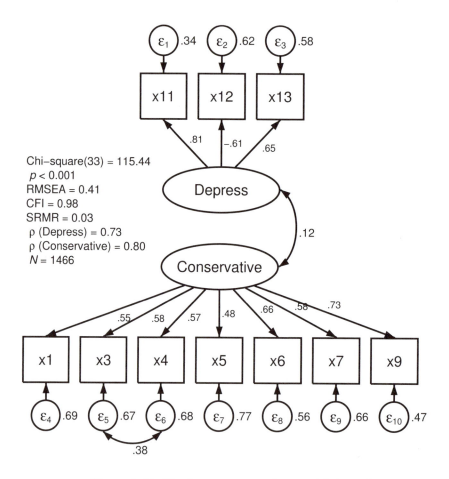

Figure 1.12. Final presentation figure with legend

If there is a problem fitting the legend on the figure, you can always include this information in a footnote to the figure.

You can do a lot more with the SEM Builder; I give you just a taste here. Click on **Estimation > Estimate...**. The **Model** tab provides three options for estimating parameters. You will want to use the *Maximum likelihood with missing values* option to use all available data. The *Asymptotic distribution free* option should be used when you have variables with far from normal distributions. Under the **if/in** tab, you can enter restrictions on your sample like we did in the text (see ch1.do). The **Weights** tab allows you to select sampling weights when these are available for your data. The **SE/Robust** tab provides many options for how you estimate the standard errors. Clicking on *Survey data estimation* changes your options in ways that are relevant for working with complex samples. Without clicking on *Survey data estimation*, you have a wide variety of estimators for the standard errors. The `Clustered robust` estimator adjusts for lack of independence within clusters, like you might get if you sampled 50 children from each of 20 different schools, with the school being your cluster. The `Robust` estimator uses the sandwich estimator to compute standard errors and thereby avoids the assumption of a normal distribution. The `Bootstrap` estimator of standard errors was illustrated in the text of this chapter, but it can also be used from here.

When you click on **Estimation > Goodness of fit**, you get fit options that are available as postestimation commands that were illustrated in the chapter text. These include the equation-level goodness of fit (`estat eqgof`) and the overall goodness of fit (`estat gof`) commands. The modification indices (`estat mindices`) are accessed by clicking on **Estimation > Testing and CIs > Modification indices**. You also get many more options that go beyond the scope of this chapter. When you run these options, the results appear in the Results window rather than on the figure.

Stata uses a general number format with up to seven total digits by default when reporting the path coefficients. If a path has 0 for its second digit—for example, 0.50—Stata reports only the first digit, .5. You may want to change this to report two decimal places. To do this, you would click on **Settings > Connections > Paths...** to open the dialog box. In the lower left on the **Results** tab, click on the button labeled **Result 1...**. Change the *Format* to %4.2f.

In like manner, we can change the format for covariance between `Depress` and `Conservative`, which was .035 for the unstandardized solution. Click on **View** and deselect **Standardized Estimates**. When you do this, the unstandardized solution replaces the standardized solution on your figure. Now click **Settings > Connections > Covariances...**. On the **Results** tab, click the button labeled **Result 1...** and change the *Format* to %4.2f. That format changes the .035 to 0.04.

I will illustrate how to use SEM Builder again in each of the following chapters. Try new features as you learn them. You will find this to be a remarkable tool that Stata has built. It works as well as or better than specialized commercial drawing programs, and they cannot even fit your model.

2 Using structural equation modeling for path models

2.1 Introduction

In this chapter, we will learn about the second building block in structural equation modeling. Now that you have some experience with the measurement model—that is, confirmatory factor analysis—we will see how to fit path models by using the `sem` command. We will finally delve into full structural equation modeling in chapter 3, which will combine latent variables presented in chapter 1 with the ideas of path analysis presented in this chapter.

The path analysis part of a structural equation model is known as the structural component, whereas the confirmatory factor analysis is known as the measurement component. If you already have learned about path analysis using standard regression-related commands, you will find that structural equation modeling both simplifies and enhances what you have been doing. If you have no experience with path analysis, you need to pay special attention to the first sections of this chapter.

2.2 Path model terminology

A path model is sometimes referred to as a causal model. You have several variables that are related in a particular fashion. When we refer to this as a causal model, we do so in a highly restricted sense. With path analysis, we often do not have randomization of participants to groups, nor do we typically have experimenter control over exposure to the independent variables. Absent randomization and control of the level of the independent variable, nearly endless alternative explanations can be offered compared with the model we propose. To illustrate how this limits the meaning of causality, we could reverse each path in most path models and get equally good results using strictly statistical criteria. This means that our causal argument must be deeply grounded in a set of strong theoretical predictions.

We can strengthen our causal argument by having the variables arranged in a time order. If X occurs before Y, then at least we can argue that Y does not cause X. However, this does not necessarily mean that X causes Y because there are endless other variables occurring before X that may cause both X and Y.

Suppose that some variable Z causes both X and Y. In our model explaining Y, the true model would be

$$Y_i = a + B_1 X_i + B_2 Z_i + e_i$$

But, let us say you do not include Z because you do not have it measured or you did not think about it. The model you fit is

$$Y_i = a + B_1' X_i + e_i'$$
$$\text{where } e_i' = B_2 Z_i + e_i$$

In this case, if the correlation between X and Z is greater than 0, then X will be correlated with e_i', resulting in a biased estimate of B_1, the direct effect of X on Y. This is one example of omitted variable bias, which is sometimes referred to as an endogeneity problem. We never know whether all relevant predictors have been included in our model, and one of those we left out may actually cause the endogenous outcome variable. Thus even with longitudinal data, we have extremely limited meaning for the term "causal model". Many researchers say that "X influences Y" or "X is associated with Y" instead of asserting that "X causes Y". A discussion of what is meant by a causal model is beyond the scope of this chapter, but it is an important issue that every researcher should examine (see Bollen and Pearl [2013]; Halpern and Pearl [2005]; Hedström and Swedberg [1998]; Shadish, Cook, and Campbell [2002]).

2.2.1 Exogenous predictor, endogenous outcome, and endogenous mediator variables

Path models typically contain three types of variables: exogenous variables, endogenous outcome variables, and endogenous mediator variables.

Exogenous variables are not causally dependent on any other variables in your model, and you are not going to try to explain them. In a traditional multiple regression, these are the independent variables. They may be correlated with one another, but you do not specify any causal direction between them. Although all exogenous variables are independent variables, not all independent variables are exogenous variables. You may also find that an exogenous variable in one model is not exogenous in another model. If you are explaining a person's happiness, you might think of education as an exogenous variable that leads to greater happiness. On the other hand, you might be interested in explaining how much education a person achieves as dependent on background characteristics, such as the education level of that person's parents; in this case, the person's education would not be an exogenous variable. Often, we have background characteristics (gender, ethnicity, education) that serve as exogenous variables. I think of the first two letters of exogenous—that is, ex—as asserting that any explanation of these variables is external to your path model.

Both types of endogenous variables are variables that are explained by your model. An endogenous outcome variable is a dependent variable with respect to all other variables in your model. For example, if your ultimate goal is trying to explain how much education a person achieves, then their education level would be the endogenous outcome. You are not interested in what the person's educational achievement explains, just in explaining the person's educational achievement. If you are trying to explain why some people follow physician instructions more closely than other people, then compliance is the endogenous outcome variable. (In such a model, a patient's education might be an exogenous variable.)

An endogenous mediator variable is independent with respect to some variables in your model and dependent with respect to other variables. The mediator variable intervenes between an exogenous variable and an endogenous outcome variable. It may also intervene between two other endogenous variables in a complex model. Endogenous mediator variables are often of great theoretical importance because they help us explain the relationship between an exogenous variable and an endogenous outcome variable. Mediator variables provide the causal mechanism linking the exogenous variable to the endogenous outcome variable.

2.2.2 A hypothetical path model

Figure 2.1 shows a path model between a series of hypothetical variables. The model has six variables labeled x1–x6. There are two exogenous variables: x1 and x2. The model explains neither of these variables; they are strictly independent variables with respect to x3–x6. The curved line with an arrow at both ends that connects the exogenous variables allows for x1 and x2 to be correlated. We do not have a direct effect from x1 → x2, nor from x2 → x1. We are not explaining either of these exogenous variables, but we are acknowledging that they may be correlated.

The only endogenous outcome variable in figure 2.1 is x6, the final outcome. The remaining variables, x3–x5, are endogenous variables that mediate some part of the effect of antecedent variables on subsequent variables. Each of these endogenous variables is explained by other variables in the model. We would say that x1 has a direct effect on x6, β_{61} (x1 → x6).[1]

---

1. Notice the order of the subscripts on β_{61}. The first is the dependent variable, x6, and the second is the independent variable, x1. When you write out a simple regression equation, such as $Y = a + BX + e$, your dependent variable is on the left and your independent variable is on the right; thus the reason that the x6 variable is the subscript on the left while the x1 is on the right.

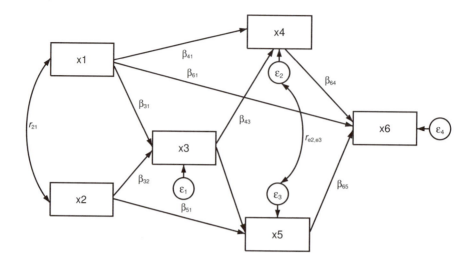

Figure 2.1. Recursive path model

We say that x4 is an endogenous variable because the model explains it; that is, x4 is caused by both x1 and x3, with coefficients β_{41} and β_{43} in our model. We also say that x4 is an endogenous mediator variable because it mediates part of the effect of x1 on x6. I do not show this algebraically, but it can be shown that the indirect effect is simply the product of the direct effects. Thus an indirect effect of x1 on x6 that is mediated by x4 is the product of the direct effect of x1 on x4, β_{41}, times the direct effect of x4 on x6, β_{64}. The indirect effect $\beta_{41} \times \beta_{64}$ reflects the indirect path x1 $\rightarrow$ x4 $\rightarrow$ x6. Notice that x1 has both a direct and an indirect effect on x6. By contrast, x3 has no direct effect on x6, but it does have two indirect effects: $\beta_{43} \times \beta_{64}$ reflecting x3 $\rightarrow$ x4 $\rightarrow$ x6 and $\beta_{53} \times \beta_{65}$ reflecting x3 $\rightarrow$ x5 $\rightarrow$ x6.

In figure 2.1, we have error terms ϵ_1–ϵ_4 for our endogenous variables x3–x6. The error term is the unexplained variance in the variable. We do not have error terms for x1 and x2 because they are both exogenous; only endogenous variables have these error terms.

Pay special attention to the curved line connecting ϵ_2 and ϵ_3. Although there is no direct causal link between x4 and x5, we are assuming that some of the unexplained variance in x4 will be correlated with some of the unexplained variance in x5. You need to consider this possible correlated residual whenever you have endogenous variables (both outcome and mediator) that are not causally related to one another. Without allowing for correlated residuals, we would be assuming that all the covariance between both variables was completely explained by their antecedent variables.[2]

---

2. Without structural equation modeling, you would need to use special regression commands, such as seemingly unrelated regression (**sureg**), when you have correlated residuals. It would be inappropriate to fit this model using Stata's **regress** command.

Figure 2.1 is a recursive path model because the flow of influence goes in a single direction; there is no feedback. We say that x1 directly, with effect β_{61}, and indirectly, with effects $\beta_{41} \times \beta_{64}$, $\beta_{31} \times \beta_{43} \times \beta_{64}$, and $\beta_{31} \times \beta_{53} \times \beta_{65}$ causes the level of x6. We do not allow for x6 to have any feedback to x1. Study figure 2.1 carefully to see if you understand how these indirect influences are simply calculated by multiplying a combination of direct effects. Can you figure out the direct and indirect effects of x2 on x6?[3]

This is a recursive model because we can arrange the variables so that there is no feedback. Whenever there is feedback, the model is called nonrecursive. Many everyday applications will have feedback. A study of the relationship between physicians and patients might include a variable for the physician's positive affect toward the patient and another variable for the patient's positive affect toward the physician. If the physician likes the patient, this might be reflected back, leading the patient to like the physician more. Likewise, the more the patient likes the physician may be reflected back, leading the physician to like the patient more. This feedback in a nonrecursive model makes the relationship circular and more difficult to estimate.

The circularity in a nonrecursive model can also occur in other ways. In figure 2.1, x3 does not directly influence x6 but does so indirectly. Thus assuming all the coefficients are positive, if you have a higher score on x3, you are expected to have higher scores on x4 and x5 and therefore on x6. If we inserted a direct path from x6 back to x3, β_{36} (x6 $\rightarrow$ x3), the model would become circular and therefore nonrecursive. In chapter 1, we discussed identification of a confirmatory factor analysis. We do not need to worry about identification for a path model as long as the model is recursive; all recursive path models are identified. Many researchers use a recursive model even when an argument could be made for some feedback, simply because it is much more difficult to identify nonrecursive models.

---

3. There is no direct effect, but there are three indirect effects: $\beta_{52} \times \beta_{65}$, $\beta_{32} \times \beta_{53} \times \beta_{65}$, and $\beta_{32} \times \beta_{43} \times \beta_{64}$.

Box 2.1. Traditional approach to mediation

Many of the early applications of mediation used a three-variable model where x was the exogenous variable, m was the endogenous mediator, and y was the endogenous outcome. They had a series of standardized steps for evaluating these models.

First, we regress y on x: **regress y x, beta.** This yields a significant value β (labeled **c**). Second, we regress m on x: **regress m x, beta.** This yields another significant β (labeled **a**). Third, we regress y on both x and m: **regress y x m, beta.** This produces a β for m (labeled **b**) and a β for x (labeled **c'**). Thus **c'** is the direct effect of x on y controlling for the mediator. If **c'** is small and insignificant, then we can say that m fully mediates the effect of x on y. If **c'** is significant but smaller than **c**, we can say m partially mediates the effect of x on y. There are special though rare cases in which **c** may not be significant, but x still indirectly influences y through the mediator. Traditionally, if x and y were not significantly correlated, then they would stop there—assuming there was no effect to be mediated.

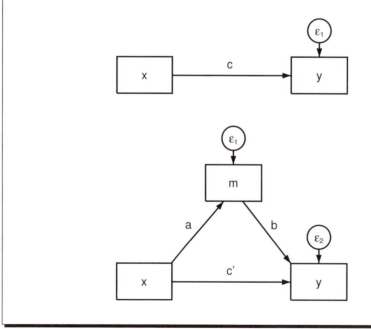

Box 2.1. (*continued*)

Stata's `regress` command reports the significance of the unstandardized coefficients, but it does not report the significance of the standardized coefficients. The significance of the standardized coefficients will not necessarily be the same as that of the unstandardized coefficients. The standardized coefficient is equal to the unstandardized coefficient times the ratio of the standard deviation of the predictor to the standard deviation of the outcome: $\beta = B \times (\text{SD}x/\text{SD}y)$. Its standard error needs to account for the variability of the two standard deviation estimates. The practical matter is that most researchers conventionally report only the significance of the unstandardized estimates even when they are reporting only the standardized coefficients.

Today, many path models are like our figure 2.1 in the sense that there are multiple endogenous mediators. Some of these may result in positive indirect effects, and some may result in negative indirect effects.

Consider a hypothetical study of married women who are members of conservative Christian churches. We will say x is how much a woman subscribes to her church doctrine, m1 is how committed she is to her marriage, m2 is how egalitarian her relationship is with her husband, and y is her marital satisfaction. Let us assume that the correlation between x and y is 0.0. Does this mean there is no mediation? Not necessarily. The women who subscribe the most to conservative church doctrine (x) may be more committed to their marriage (m1), and this may in turn lead to greater marital satisfaction (y). This would represent a positive indirect effect of x on y mediated by m1. By contrast, greater endorsement of conservative church doctrine (x) may lead to a less egalitarian relationship with her husband (m2), which could lead to lower marital satisfaction (y). This represents a negative indirect effect of x on y mediated by m2. When you have positive and negative indirect effects, these may offset each other and result in a correlation of x and y of 0.0. There is still important mediation going on, it is just more complicated than traditional models with just a single mediator.

2.3 A substantive example of a path model

Suppose you believe that children with better attention span skills at the age of 4 are going to have advantage in their academic development. How far could this go? A child with a strong attention span at age 4 might be able to read better and perform math better at age 7. We could speculate that the attention span at age 4 would carry all the way over to math skills as a young adult, say, age 21. We might further speculate that some of this could be a persistent direct effect and some could be an indirect effect mediated by how well the child does at reading and math at age 7. This example uses data from McClelland et al. (2013).

How would we represent such a model? Figure 2.2 shows an initial model that ignores the possible correlation between ϵ_1 and ϵ_2.

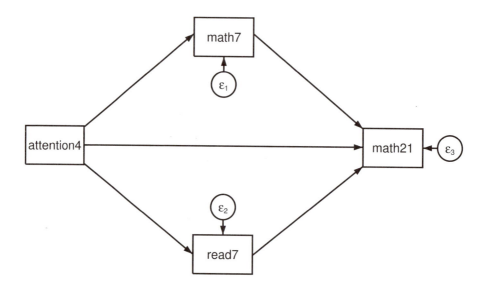

Figure 2.2. Attention span at age 4 and math achievement at age 21 ($N = 430$)

Let us fit this model just as it is drawn. It is always useful to draw a figure before writing the command, even if the drawing is just a freehand drawing using a pencil. Figure 2.2 was drawn using Stata's SEM Builder (see appendix A and the appendix to this chapter). The SEM Builder allows us to estimate the results directly from the figure. You may want to rely on the SEM Builder as your primary way of fitting models; however, in the text of this chapter, I show the actual Stata commands. It is important to understand what these commands are, even if you rely on the SEM Builder most of the time.

The `sem` command needs to estimate each of the structural paths in figure 2.2. We do not need to specify the paths from the error (residual) terms to their correspondent variables because Stata assumes you have some error to represent the unexplained variance. There is no measurement model in this example because there are no latent exogenous or latent endogenous variables in figure 2.2; all the variables are directly observed measures. Here is the set of Stata commands we run:

```
. sem (math7 <- attention4)              /// Equation for math7
      (read7 <- attention4)              /// Equation for read7
      (math21 <- attention4 math7 read7),  /// Equation for math21
      method(mlmv) standardized          /// Missing values and standardized
                                         //  solution
. estat eqgof                            //  Provide equation R-squares
. estat gof, stats(all)                  //  Provide measures of goodness of fit
. estat teffects, standardized           //  Provide indirect and total effects
. estat mindices                         //  Provide modification indices
```

The `sem` command estimates the path coefficients in the equation for each endogenous variable, `math7`, `read7`, and `math21`. The first line of the `sem` command only has `attention4` as a predictor of `math7`, which is consistent with figure 2.2. The second line does the same for `read7`. The third line predicts `math21` using all three antecedent variables, `attention4`, `math7`, and `read7`. In our `sem` command, we used a special option, `method(mlmv)`, which provides a maximum likelihood estimate using all observed values. It is designed explicitly for situations where we have missing values. If we had not used this option, the default maximum likelihood estimator would have resulted in a sample of $N = 338$ because it uses listwise deletion. By contrast, the `method(mlmv)` option uses the information from all the observations that have observed values on at least some of the variables and yields $N = 430$. Finally, the `sem` command asks for the standardized solution. Both the estimation method and the choice of a standardized solution come as options separated from the main command by a comma.

We have used four of the many possible postestimation commands available in Stata. `estat eqgof` (extended statistics reporting the equation's goodness of fit) will tell us how much variance we explain for each endogenous variable, that is, the R^2's for `math7`, `read7`, and `math21`. `estat gof, stats(all)` provides information about how well our model fits the data. This goodness-of-fit information is not a standard part of traditional regression solutions. `estat teffects, standardized` gives us measures of the direct, indirect, and total effects of variables. Finally, `estat mindices` (extended statistics reporting modification indices) highlights any places where our model does not fit the data—what we might want to change about our model to obtain a better fit.

Here is a portion of the output from the `sem` command:

```
(output omitted)
Structural equation model                        Number of obs    =        430
Estimation method  = mlmv
Log likelihood     =  -4246.557
```

Standardized	Coef.	OIM Std. Err.	z	P>|z|	[95% Conf. Interval]	
Structural						
math7 <-						
attention4	.141458	.0486307	2.91	0.004	.0461437	.2367723
_cons	3.04888	.3344304	9.12	0.000	2.393408	3.704351
read7 <-						
attention4	.1289838	.0491968	2.62	0.009	.0325598	.2254077
_cons	3.163475	.3383925	9.35	0.000	2.500238	3.826712
math21 <-						
math7	.3075685	.0481426	6.39	0.000	.2132108	.4019262
read7	.2520422	.0489132	5.15	0.000	.156174	.3479104
attention4	.1171187	.0467622	2.50	0.012	.0254664	.208771
_cons	1.380531	.361878	3.81	0.000	.6712636	2.089799
var(e.math7)	.9799896	.0137584			.9533913	1.00733
var(e.read7)	.9833632	.0126912			.9588009	1.008555
var(e.math21)	.8075246	.0341705			.7432537	.8773531

```
LR test of model vs. saturated: chi2(1)    =      27.56, Prob > chi2 = 0.0000
```

If you first draw a model like the one in figure 2.2 using the SEM Builder, you can fit the model by clicking on **Estimation > Estimate...**, on the **Model** tab choosing Maximum likelihood with missing values, and then clicking on **OK**.[4] You get the unstandardized results. You may then get the standardized results by selecting **View > Standardized Estimates**. We have cleaned up the figure by deleting some statistics, including variable means, intercepts, and variances, that Stata reports by default. You can learn how to draw and fit the model with the SEM Builder in the appendix to this chapter and in more detail in appendix A at the end of the book.

---

4. Remember, if you use a Mac, you first click on the gear-shaped tool in the upper right of your SEM Builder screen.

Figure 2.3 is the resulting model including the standardized estimates that Stata generates:

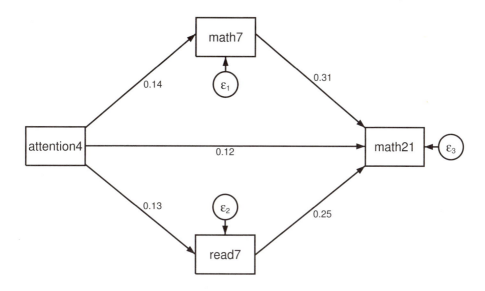

Figure 2.3. Standardized estimates for attention span at age 4 and math achievement at age 21 ($N = 430$)

The coefficients for the standardized solution are known as standardized path coefficients (sometimes referred to as standardized beta weights). For `math7`, we have just one predictor, `attention4`, and it has a standardized path coefficient of $\beta_{21} = 0.14$ ($z = 2.91$, $p < 0.01$) as shown on figure 2.3 and in the results above. The results also provide a confidence interval for each standardized parameter. A child's attention span at age 4 has a small but statistically significant effect on the child's math performance at age 7. For `read7`, the effect of `attention4` is $\beta_{31} = 0.13$ ($z = 2.62$, $p < 0.01$). We also have the effects of each variable on our endogenous outcome variable, `math21`. We see that the strongest predictor is `math7`, $\beta_{42} = 0.31$ ($z = 6.39$, $p < 0.001$), followed closely by `read7`, $\beta_{43} = 0.25$ ($z = 5.15$, $p < 0.001$). Remarkably, attention span at age 4 has a small but statistically significant direct effect on math at age 21, $\beta_{41} = 0.12$ ($z = 2.50$, $p = 0.01$), controlling for both `read7` and `math7`.

How much variance do we explain for each endogenous variable? Our second command, `estat eqgof`, gives us those results. We do not explain much variance in `math7` or `read7`—2.0% and 1.7%, respectively (see below)—but we do explain more variance in `math21`, $R^2 = 0.19$.

$$R^2 = \frac{\text{Variance predicted}}{\text{Total variance to be fit}}$$

where variance predicted is our explained variance and total variance to be fit is our total variance in the endogenous variable. This result also reports the estimated Bentler–Raykov R^2 that is appropriate when you have a nonrecursive model.

```
. estat eqgof
Equation-level goodness of fit
```

| | | Variance | | | | |
depvars	fitted	predicted	residual	R-squared	mc	mc2
observed						
math7	7.621122	.1525015	7.46862	.0200104	.141458	.0200104
read7	64.70388	1.076467	63.62742	.0166368	.1289838	.0166368
math21	6.920939	1.33211	5.588828	.1924754	.4387202	.1924754
overall				.0515245		

```
mc  = correlation between depvar and its prediction
mc2 = mc^2 is the Bentler-Raykov squared multiple correlation coefficient
```

The next command tells us how well our model fits the data. This includes some information to evaluate our model that is unavailable with traditional regression results.

```
. estat gof, stats(all)
```

Fit statistic	Value	Description
Likelihood ratio		
chi2_ms(1)	27.561	model vs. saturated
p > chi2	0.000	
chi2_bs(6)	130.877	baseline vs. saturated
p > chi2	0.000	
Population error		
RMSEA	0.249	Root mean squared error of approximation
90% CI, lower bound	0.174	
upper bound	0.332	
pclose	0.000	Probability RMSEA <= 0.05
Information criteria		
AIC	8515.114	Akaike's information criterion
BIC	8559.816	Bayesian information criterion
Baseline comparison		
CFI	0.787	Comparative fit index
TLI	-0.276	Tucker-Lewis index
Size of residuals		
CD	0.052	Coefficient of determination

```
Note: SRMR is not reported because of missing values.
```

These results are disappointing. Our model fails significantly to reproduce the co-variance matrix for our four variables, $\chi^2(1) = 27.56$, $p < 0.001$. Remember from chapter 1 that with structural equation modeling, a significant chi-squared means that we fail to account for the covariances among our variables. We want a small chi-squared relative to the degrees of freedom and one that is not statistically significant. The root mean squared error of approximation (RMSEA) is 0.25, much greater than our ideal standard of less than or equal to 0.05; the comparative fit index (CFI) is 0.79, far below our ideal standard of 0.95 or even the acceptable standard of 0.90. We have 1 degree of freedom. A recursive path model is always identified or over-identified.

We will ignore the direct and indirect effects until we get a better fitting model. Here we will examine the last command, `estat mindices`, to see if there is a way to improve the fit. This command reports modification indices, each of which is an estimate of how much we would reduce chi-squared if we added the indicated path. For example, `math7 <- read7` has a modification index of 26.89. This means that if we added a path going from `read7` to `math7`, our chi-squared would be reduced by roughly 26.89 points and this would use up 1 degree of freedom. Because a chi-squared of 3.84 is significant with 1 degree of freedom, this would be a significant improvement in how well our model fits the data. We need to be very cautious here: we have only 1 degree of freedom, and estimating any additional parameter will guarantee a perfect fit because we will then have 0 degrees of freedom. Any model with 0 degrees of freedom will have $\chi^2 = 0$, RMSEA $= 0$, and CFI $= 1.0$. That does not mean it is a good model, just that we cannot test it unless it is over-identified (that is, has at least 1 degree of freedom).

Here are the results of the `estat mindices` command:

```
. estat mindices

Modification indices
```

		MI	df	P>MI	EPC	Standard EPC
Structural						
math7 <-						
	read7	26.885	1	0.00	.0899778	.2621748
	math21	26.885	1	0.00	1.091552	1.040202
read7 <-						
	math7	26.885	1	0.00	.7665476	.2630773
	math21	26.885	1	0.00	2.615316	.8553455
cov(e.math7,e.read7)		26.885	1	0.00	5.725053	.2626257

```
EPC = expected parameter change
```

These estimates are one-at-a-time values. Each additional parameter we estimate uses 1 degree of freedom.

Some of the possibilities make no sense. For example, take `math21` → `math7` and `math21` → `read7`: they would require an effect to go backward in time. By contrast, `read7` → `math7` might make sense: if the math test included any thought problems, reading skills might be helpful. It is also possible that `math7` → `read7`, but this would be harder to justify.

At the bottom of the results, the final extra path we might add would be to allow the error terms for `math7` and `read7`—that is, `e.math7` and `e.read7`—to be correlated. In figure 2.3, `e.math7` appears as ϵ_1 and `e.read7` appears as ϵ_2. Allowing these error terms to be correlated makes a lot of sense, and we do not have to make a causal argument as we would for `math7` → `read7` or vice versa. This correlated error means that there are variables that would influence both `math7` and `read7` that are not in our model. There are lots of these, such as the child's socioeconomic status and gender. Allowing the error terms of endogenous variables that have no direct linkage between them to be correlated is similar to the idea of a partial correlation. It is how much of the variance in `read7` that is unexplained by our model is correlated with the variance in `math7` that is unexplained by our model. Because we have not explained about 98% of the variance of both variables (R^2's for both variables were about 0.02), it is likely that there will be some covariance.

2.4   Estimating a model with correlated residuals

Allowing `e.math7` and `e.read7` to be correlated will necessarily make our model fit perfectly because it will reduce our degrees of freedom to 0. We will add this link, however, because it makes sense for it to be there. As a generalization, when you have parallel mediators (`read7` and `math7`) that both mediate some of the effect of attention span at age 4 on math performance at 21, it is usually important to allow their errors to be correlated. Figure 2.4 presents our revised model:

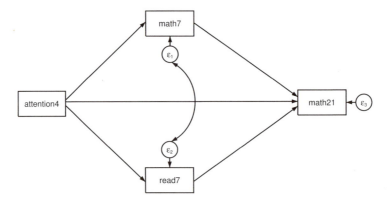

Figure 2.4. Revised model of attention span at age 4 and math achievement at age 21 with correlated errors ($N = 430$)

To fit this model, we run exactly the same set of commands as before except we additionally request the covariance of ϵ_1 and ϵ_2. We add this using `covariance()`, abbreviated `cov()`, as an additional option after the comma:

```
. sem (math7 <- attention4)                      /// Equation for math7
      (read7 <- attention4)                      /// Equation for read7
      (math21 <- attention4 math7 read7),        /// Equation for math21
      method(mlmv) cov(e.math7*e.read7) standardized //  Missing values, covariance
. estat eqgof                                    //  Provide R-square
. estat gof, stats(all)                          //  Goodness-of-fit measures
. estat teffects, standardized                   /// Standardized indirect
                                                 //  effects
. estat mindices                                 //  Modification indices
```

The results of the `sem` command (results not shown) are almost the same as before, except we now find a covariance of `e.math7` and `e.read7`. Because we requested the standardized solution, the reported value is actually a correlation, $r_{\text{e.math7,e.read7}} = 0.26$ ($z = 5.54$, $p < 0.001$). I leave it to you to run this structural equation model using the commands or using the SEM Builder. The R^2's for `read7` and `math7` have not changed much compared with what we had in the initial model, but the R^2 for `math21` has increased a bit from 0.19 to 0.22.

```
. estat eqgof
Equation-level goodness of fit
```

depvars	fitted	Variance predicted	residual	R-squared	mc	mc2
observed						
math7	7.619575	.1546551	7.46492	.0202971	.1424678	.0202971
read7	64.62929	1.086548	63.54274	.016812	.1296611	.016812
math21	7.178428	1.593784	5.584644	.2220241	.4711943	.2220241
overall				.0448891		

```
mc  = correlation between depvar and its prediction
mc2 = mc^2 is the Bentler-Raykov squared multiple correlation coefficient
```

Many of the statistics produced by the `estat gof, stats(all)` command (results not shown) are meaningless because we have no degrees of freedom and will have a perfect fit, by definition. The `estat mindices` command reports no modification indices because we have a perfect fit, as we must have with no degrees of freedom.

2.4.1 Estimating direct, indirect, and total effects

We have not yet discussed `estat teffects, standardized`, which provides the direct, indirect, and total effects[5] for each predictor. In our model, we have three predictors: `attention4` predicts `read7`, `math7`, and `math21`; `read7` predicts `math21`; and `math7` predicts `math21`. The command provides the following three tables of results:

---

5. Here total effects = direct effects + indirect effects.

```
. estat teffects, standardize
```

Direct effects

		OIM			
	Coef.	Std. Err.	z	P>|z|	Std. Coef.
Structural					
math7 <-					
attention4	.1290411	.0446933	2.89	0.004	.1424678
read7 <-					
attention4	.342035	.1313072	2.60	0.009	.1296611
math21 <-					
math7	.2920135	.0470959	6.20	0.000	.3008525
read7	.0820604	.0161567	5.08	0.000	.2462258
attention4	.1009146	.0408745	2.47	0.014	.1147871

Indirect effects

		OIM			
	Coef.	Std. Err.	z	P>|z|	Std. Coef.
Structural					
math7 <-					
attention4	0	(no path)			0
read7 <-					
attention4	0	(no path)			0
math21 <-					
math7	0	(no path)			0
read7	0	(no path)			0
attention4	.0657493	.0202659	3.24	0.001	.0747877

Total effects

		OIM			
	Coef.	Std. Err.	z	P>|z|	Std. Coef.
Structural					
math7 <-					
attention4	.1290411	.0446933	2.89	0.004	.1424678
read7 <-					
attention4	.342035	.1313072	2.60	0.009	.1296611
math21 <-					
math7	.2920135	.0470959	6.20	0.000	.3008525
read7	.0820604	.0161567	5.08	0.000	.2462258
attention4	.1666639	.0440972	3.78	0.000	.1895749

In reading the tables, notice that the standardized values appear in the far right column; the column labeled Coef. reports the unstandardized values, while the column labeled Std. Coef. reports the standardized values.

The first table reports the unstandardized and the standardized direct effects. These are the same standardized direct effects we already had. The z test and probability reported in this table are for the unstandardized direct effects. You can either rely on the tests for the unstandardized solution or see how you can compute tests for the standardized effects by reading box 2.3.

The second table reports indirect effects, where they exist. By examining figure 2.4, we see that there is no indirect effect of `attention4` on `math7`, just a direct effect. Similarly, there is no indirect effect of `attention4` on `read7`. There are two indirect paths between `attention4` and `math21`: `attention4` $\rightarrow$ `math7` $\rightarrow$ `math21` and `attention4` $\rightarrow$ `read7` $\rightarrow$ `math21`. Stata provides the total of these indirect effects, 0.07, and a z test for the unstandardized indirect effect, $z = 3.24$, $p = 0.001$; Stata does not provide the two separate effects. We can easily calculate each of these indirect effects by hand multiplying the corresponding paths together ($\beta_{21} \times \beta_{42}$ and $\beta_{31} \times \beta_{43}$), but we would not have separate tests of significance for them. I explain how you can test these two indirect effects in box 2.3.

The last table in our results provides the total effects by adding together the direct and corresponding indirect effects (see box 2.2). The `Std. Coef.` column provides these results. The total standardized effect of `attention4` on `math21` is the direct effect plus the indirect effect, that is, $0.115 + 0.075 = 0.190$; the z test for the unstandardized total effect is $z = 3.78$, $p < 0.001$.

Let us put all this information together in both a figure and a table. In figure 2.5, we report the standardized path coefficients for the direct effects as well as the correlation of the error terms. In addition, we report the available measures of goodness of fit. With no degrees of freedom, the standard measures (chi-squared test, CFI, and RMSEA) are meaningless, so we do not report them here; We would report them for any model that had at least 1 degree of freedom. We also report the R^2 for each endogenous variable and the sample size.

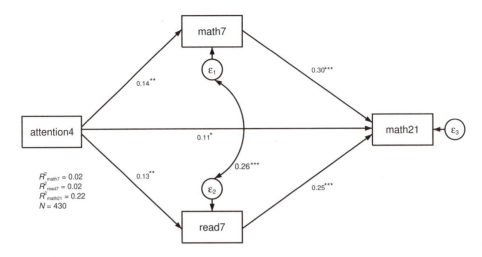

Figure 2.5. Standardized estimates for attention span at age 4 and math achievement at age 21 with correlated errors ($N = 430$; $*p < 0.05$, $**p < 0.01$, and $***p < 0.001$)

Table 2.1 shows the direct, indirect, and total effects. We can get all the information we need from the `estat teffects, standardized` command results. We report the significance levels based on the z tests for the unstandardized solution because the specific z tests for the indirect and direct effects are not provided in the standardized solution. While the test of an unstandardized coefficient and the test of the corresponding standardized coefficient may differ, this difference usually does not change the overall significance level.

Table 2.1. Standardized effects of attention span at age 4 and math achievement at age 21 with correlated residual for math at age 7 and reading at age 7

Outcome	Direct effect	Indirect effect	Total effect
Math at age 7			
attention4 → math7	0.14**[A]	–	0.14**
Reading at age 7			
attention4 → read7	0.13**	–	0.13**
Math at age 21			
attention4 → math21	0.11*	0.07**	0.19***
math7 → math21	0.30***	–	0.30***
read7 → math21	0.25***	–	0.25***

[A] The significance levels shown here are for the unstandardized solution.
* $p < 0.05$, ** $p < 0.01$, and *** $p < 0.001$.

Box 2.2. Proportional interpretation of direct and indirect effects

Many researchers present the direct and indirect effects as we do in table 2.1. It may also be important to know the proportion of the total effect that is direct and the proportion that is indirect. Is most of the effect direct? Does the indirect effect represent just a small portion of the total effect? These are important questions to answer for your reader, and the computations are quite simple.

For our standardized results, the total effect of `attention4` on `math21` is 0.190. The direct component of this total effect is 0.115, so we can say that $0.115/0.190 = 0.605$ or 60.5% of the effect of `attention4` on `math21` is direct after controlling for both `read7` and `math7`. By contrast, the indirect effect is 0.07, so that we can say that $0.075/0.190 = 0.395$ or 39.5% of the effect is indirect. Thus even after controlling for `read7` and `math7`, the majority of the effect of `attention4` on `math21` is a direct effect, and there is a sizable but smaller percentage of the effect that is indirect.

Some authors compute the proportions of direct and indirect effects with the unstandardized coefficients. This should give you the same results subject to rounding error. This is true regardless of the scales on which the unstandardized coefficients are based, because the standardized coefficient is equal to the unstandardized coefficient times the ratio of the standard deviation of the predictor to the standard deviation of the outcome, $\beta = B \times (\text{SD}x/\text{SD}y)$. By substituting the B's and the formulas for the β's into the ratios, the standard deviations cancel out (personal communication Medeiros [2012]).

Box 2.3. Testing standardized indirect and specific indirect effects

Although Stata provides the indirect effect of `attention4` on `math21`, it does not provide the specific indirect effects that contribute to this. Also, when you run `estat teffects, standardized`, the tests of significance are for the unstandardized coefficients, not for the standardized coefficients. Although many researchers report only the unstandardized tests, I explain in this box how you can produce the correct tests for the standardized coefficients.

We have two specific indirect effects: `attention4` → `math7` → `math21` and `attention4` → `read7` → `math21`. Both of these combine to produce the indirect effect reported by the `estat teffects` command. The first of these can be calculated by multiplying the coefficient on the path from `attention4` to `math7` (0.14) times the coefficient on the path from `math7` to `math21` (0.30), yielding $0.14 \times 0.30 = 0.04$. Thus the specific indirect effect of attention span at age 4 on math performance at age 21 as mediated by math performance at age 7 is 0.04. Because the total effect of `attention4` on `math21` is 0.19, we can calculate the percentage of the total effect of attention span on `math21` that is mediated through `math7` as $0.04/0.19 = 0.21$, or 21%.

The second indirect path of `attention4` on `math21` is mediated by reading at age 7. We obtain this indirect path by multiplying $0.13 \times 0.25 = 0.03$. Thus $0.03/0.19 = 0.16$, or 16% of the total effect of `attention4` on `math21` is mediated by `reading7`. Because there are two indirect effects of `attention4` on `math21`, the result in table 2.1 reports that the total indirect effect of `attention4` on `math21` is 0.075, that is, $0.043 + 0.032$. The table tells us that this combined indirect effect is significant at the $p < 0.01$ level.

We are often interested in the specific indirect effects. Does attention span have a significant indirect effect mediated by `math7`? (Is 0.04 statistically significant?) Does attention span have a significant indirect effect mediated by `read7`? (Is 0.03 statistically significant?) These require a nonlinear test because the estimated indirect effects are products of parameters. These tests are not available automatically. Performing these nonlinear tests is not difficult using the nonlinear comparison command, `nlcom`, but it is a bit tedious and requires attention to labeling.

To use `nlcom`, we first make up a name for each indirect effect so that we can find the estimated values in the results. First, let us name `attention4` → `math7` → `math21` as `attn_m7_m21`. We will name `attention4` → `read7` → `math21` as `attn_r7_m21`.

Box 2.3. (*continued*)

Next we need to find out the names Stata has given to the coefficients for each of the direct paths in the model. We need a legend of the names of coefficients, which we can get by replaying the results with `sem, coeflegend`.

Here is part of the legend:

```
. sem, coeflegend
```

(*output omitted*)

	Coef.	Legend
Structural		
math7 <-		
attention4	.1290411	_b[math7:attention4]
_cons	8.399792	_b[math7:_cons]
read7 <-		
attention4	.342035	_b[read7:attention4]
_cons	25.42185	_b[read7:_cons]
math21 <-		
math7	.2920135	_b[math21:math7]
read7	.0820604	_b[math21:read7]
attention4	.1009146	_b[math21:attention4]
_cons	3.658488	_b[math21:_cons]

(*output omitted*)

This command reports the unstandardized coefficients, but do not let that worry you now. We are only interested in how we should refer to the coefficients; for example, the coefficient corresponding to the direct effect of `attention4` on `math7` is referred to as `_b[math7: attention4]`. The command below computes the specific indirect effects by multiplying the direct effects. When we run the `nlcom` command, we need to add the prefix command `estat stdize:` so that Stata uses the standardized solution.

Box 2.3. (*continued*)

Here is our full command and the results of the nonlinear combinations test:

```
. estat stdize: nlcom                              // Standardized
> (attn_m7_m21: _b[math7:attention4]*_b[math21:math7]) // Nonlinear tests
> (attn_r7_m7: _b[read7:attention4]*_b[math21:read7])

  attn_m7_m21:  _b[math7:attention4]*_b[math21:math7]
  attn_r7_m7:   _b[read7:attention4]*_b[math21:read7]
```

	Coef.	Std. Err.	z	P>|z|	[95% Conf. Interval]	
attn_m7_m21	.0428618	.0161643	2.65	0.008	.0111803	.0745433
attn_r7_m7	.0319259	.0137121	2.33	0.020	.0050507	.0588011

The indirect effect through `math7`, labeled `atten_m7_m21`, is 0.04, as we calculated by hand. Now we also have a test of significance for it and a confidence interval. In writing a report, we would write that attention span at age 4 has a small but statistically significant standardized indirect effect on math skills at age 21 that is mediated by math skills at age 7 of 0.04, $z = 2.65$, $p < 0.01$. Attention span at age 4 also has a small but statistically significant indirect effect on math skills at age 21 that is mediated by reading skills at age 7 of 0.03, $z = 2.33$, $p = 0.02$.

Here is a summary of all the commands we ran in this analysis:

```
. sem (math7 <- attention4)                  /// Eq. for math7
  (read7 <- attention4)                       /// Eq. for read7
  (math21 <- attention4 math7 read7),         /// Eq. for math21
  method(mlmv) cov(e.math7*e.read7)           /// Missing, covariance
  standard                                    // Standardized

. sem, coeflegend                             // Coefficient legend

. estat stdize:                               // Stand. nonlinear tests
  nlcom (attn_m7_m21: _b[math7: attention4] *_b[math21: math7])
  (attn_r7_m7: _b[read7: attention4] *_b[math21: read7])

. estat teffects, nodirect nototal standardize  // Stand. indirect
```

2.4.2 Strengthening our path model and adding covariates

In the introduction to this chapter, I stressed the limited meaning of causality when people refer to path models as causal models. There are things we can do to strengthen this meaning. The worse justification for a causal argument is when you have all your data collected at one time point.

Suppose we had measured attention span at age 7 and reading at age 7. There is no time ordering to help us know which variable is the cause and which is the effect. Do children who have good attention spans at age 7 therefore read better (`attention7` → `read7`)? Do children who read a lot at age 7 develop better attention spans (`read7` → `attention7`)? Do these two variables simultaneously influence each other (`attention7` ↔ `read7`)? It is impossible to make a convincing argument that the causal direction is one way or the other.

In our example, we measured attention span at age 4 and reading at age 7, so we can safely make the argument that the causal direction is `attention4` → `read7`. Is our time ordering enough for a strong causal interpretation? Not really. In experimental research, we would have randomized trials where the researcher randomly assigns people to conditions and then controls their level of exposure to the independent variable. Without the random assignment, there could be many other covariates that we have not included in our model.

Perhaps gender is related to both the attention span at age 4 and the reading level at age 7 (`gender` → `attention4` and `gender` → `read7`). When there is a common antecedent that causes both variables—that is, gender causes both attention span at 4 and reading at 7—we say the relationship between attention span at 4 and reading at 7 is spurious or at least partially spurious.

Alternatively, other covariates might contribute to reading ability at age 7. Perhaps a child's vocabulary when he or she is 4 leads to better reading at age 7. Figure 2.6 illustrates both types of covariates. If we control for gender, then the relationship between `attention4` and `read7` may be weakened because the relationship is spurious or at least partially spurious. If we control for the child's vocabulary at age 4, that may diminish the strength of the relationship between `attention4` and `read7`.

We can control for covariates by using the `sem` command. We do this by adding the covariates as predictors of the endogenous variables we think they directly influence. In figure 2.6, we added paths from gender to both attention span at 4 and reading at 7, and a path from vocabulary at 4 to reading at 7. We also show a correlation between `gender` and `vocab4`.[6]

---

6. This is not necessary because Stata assumes these exogenous variables are freely correlated; however, it is sometimes nice to include this for readers who are unfamiliar with Stata. When you have several exogenous variables, including all the correlations becomes very messy and detracts from the main point of your figure. In that case, leave them out of the figure and add a footnote indicating that the exogenous variables were allowed to be correlated.

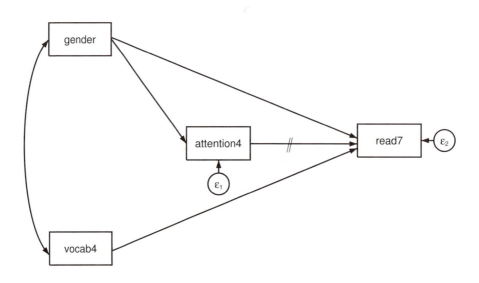

Figure 2.6. Controlling for covariates can weaken a relationship, as illustrated by gender and vocabulary skills at age 4

Let us make this a bit more complicated. We will focus on the model in figure 2.5 except we will add four covariates: vocabulary at 4, adoption status of the child, gender of the child, and his or her mother's education.

Let us look at the commands:

```
. sem (math7 <- attention4 vocab4 adopted male momed)             /// Eq. for math7
    (read7 <- attention4 vocab4 adopted male momed)               /// Eq. for read7
    (math21 <- attention4 vocab4 adopted male momed math7 read7), /// Eq. for math21
    cov(e.math7*e.read7) standardized method(mlmv)                //  Cov. errors
. estat eqgof                                                     //  R-squared
. estat gof, stats(all)                                          /// Goodness of
                                                                  //  fit
. estat teffects, standardize                                    /// Indirect
                                                                  //  effect
. estat mindices                                                 //  Modification
```

We have added four covariates for each equation: `vocab4`, `adopted`, `male`, and `momed`. In the first line, the equation for `math7`, we are now saying that the `math7` score may be influenced by `attention4` as in our original model, but also by the child's vocabulary at age 4, whether he or she is adopted, whether the child is a boy or girl, and his or her mother's education. When we fit this equation, the question becomes: Does `attention4` still have a significant effect on `math7`? We do the same thing in the next two lines for `read7` and `math21`.

When reporting our results, we still have 0 degrees of freedom because we are adding as many new parameters to be estimated as we have added variances and covariances. Therefore, we still do not have a test of fit for the model. We would present the results as we did in figure 2.5 and table 2.1; we would simply add a footnote to the figure and the table stating that we have controlled for vocabulary at age 4, adoption status, gender, and mother's education. Because these are simply covariates we want to control, we do not need to include their parameter estimates in the figure or table.

Adding covariates in this way makes a stronger argument. Even when we control for these four covariates, the attention span a child has at age 4 still has a statistically significant direct effect on his or her reading and math performance at age 7 as well as statistically significant direct and indirect effects on his or her math performance at age 21 (results not shown). Still, we need to emphasize that we have not made as strong a case for a causal effect as we would with a randomized trial. There simply are too many other possible covariates that we have not included in our model. Unless we included all possible covariates, we could not rule out the possibility that the relationship between attention and math/reading is noncausal.

2.5 Auxiliary variables

Missing values are often a problem. In our example that includes covariates, had we used the default maximum likelihood estimation, which uses listwise deletion, we would have had 299 people. By using the `method(mlmv)` option—and thereby using all available information—we instead have 430 people.

Listwise deletion assumes the values will be missing completely at random (MCAR), although this assumption can be relaxed when estimating unstandardized structural coefficients. As its name implies, MCAR asserts that whether you have a missing value on a variable is completely random and thus missing values are not related to any variable regardless of whether the variable is in our model. This assumption is rarely justifiable.

The estimation we performed using the `method(mlmv)` option makes the more realistic assumption that missing values are missing at random (MAR). Some people consider the name MAR unfortunate because it sounds like MCAR, but these two conditions are nothing alike. MAR assumes that you have either variables in your model or additional variables that explain who answers and who does not answer the items in your dataset.

An auxiliary variable is a variable that is not part of your model but that explains who is more likely to have missing values. An example would be education because people with less education often have more missing values than people with more education. While we cannot test the MAR assumption, it is less restrictive than MCAR; sometimes we can include a few auxiliary variables that help explain who has missing values. The point to remember is that `method(mlmv)` is less restrictive than the assumption of some other methods, such as listwise (casewise) deletion, that are valid only if the missing values are MCAR. Note that `method(mlmv)` still does assume that the variables are multivariate normal, which is often a problematic assumption. When you violate the multivariate normal assumption, this can bias your parameter estimates, in some cases substantially (Acock 2012b).

Auxiliary variables that are added to make the MAR assumption more plausible either help predict a score that is missing (age might help predict a person's height if height were a missing value for that person) or help explain why variables have missing values (older people may be more likely to skip the item about their height). Although adding a few auxiliary variables helps justify the MAR assumption, they may also complicate the multivariate normal assumption, for example, using gender. Additionally, some potential auxiliary variables might have a lot of missing data themselves, which can add more noise to your model. For example, you might think of income as an auxiliary variable, but many people refuse to report their income. One solution is to fit your model using both the default maximum likelihood estimation assuming listwise deletion and the `method(mlmv)` approach. Hopefully, the results will be similar, adding confidence to your findings.

Stata's `sem` command does not have an explicit way to add auxiliary variables; however, there is a way that we can trick it. Suppose we come up with a set of six variables to serve as auxiliary variables. We will label them `aux1`–`aux6` in this example. If we can include these auxiliary variables, then we can explain the missingness and add credibility to our meeting the MAR assumption.

We simply add another equation, `(<-aux1-aux6)`, as shown below:

```
. sem (math7 <- attention4 vocab4 adopted male momed)          // Eq. for math7
      (read7 <- attention4 vocab4 adopted male momed)           // Eq. for read7
      (math21 <- attention4 vocab4 adopted male momed math7 read7) // Eq. for math21
      (<-aux1-aux6),                                            // Auxiliary
      method(mlmv)                                              //
      cov(e.math7*e.read7) standardized                        // Cov. errors
```

We are not interested in the auxiliary variables represented by `aux1`–`aux6`. By including this line in our `sem` command, these variables will provide information to help justify the MAR assumption. Because the `method(mlmv)` estimator is a full-information estimator, the information in `aux1`–`aux6` will be part of the covariance matrix we are analyzing. As a caution, we want auxiliary variables that have few missing values themselves, and we remember that adding them may add to the problem of assuming multivariate normality.

2.6   Testing equality of coefficients

By examining our results in table 2.1 or figure 2.5, we see that `math7` has a standardized direct effect of $\beta = 0.30$ on `math21`, and `read7` has a standardized direct effect of $\beta = 0.25$. In the literature, people often misinterpret the difference $(0.30 - 0.25)$ as being statistically significant. The significance we report for these two standardized path coefficients compares the coefficients to a value of 0.00, not to each other. They may both be significantly different from 0.00, but this does not ensure they are significantly different from each other. One may be significant and the other not significant, but this also does not mean they are significantly different from one another. The postestimation commands for `sem` allow us to explicitly test whether there is a significant difference between parameter estimates.

As mentioned previously, Stata stores the value of all parameter estimates in memory until a subsequent command overwrites them. To test the difference between the value of these two standardized paths, we need to know the name that Stata assigns to each of them. We get these names by replaying the results with `sem, coeflegend`, which gives us the coefficient legend.

```
. sem, coeflegend
Structural equation model                    Number of obs      =        430
Estimation method  = mlmv
Log likelihood     = -4232.7763
```

	Coef.	Legend
Structural **math7 <-**		
attention4	.1290411	_b[math7:attention4]
_cons	8.399792	_b[math7:_cons]
read7 <-		
attention4	.342035	_b[read7:attention4]
_cons	25.42185	_b[read7:_cons]
math21 <-		
math7	.2920135	_b[math21:math7]
read7	.0820604	_b[math21:read7]
attention4	.1009146	_b[math21:attention4]
_cons	3.658488	_b[math21:_cons]
var(e.math7)	7.46492	_b[var(e.math7):_cons]
var(e.read7)	63.54274	_b[var(e.read7):_cons]
var(e.math21)	5.584644	_b[var(e.math21):_cons]
cov(e.math7, e.read7)	5.662178	_b[cov(e.math7,e.read7):_cons]

```
LR test of model vs. saturated: chi2(0)    =       0.00, Prob > chi2 =      .
```

These estimates are the unstandardized estimates, but that is not going to be a problem. Now we can test the equality of standardized path coefficients by running a `test` command and using the labels shown in the legend. The `test` command gives us a Wald chi-squared test of the difference. We use the prefix command `estat stdize:` so that the test is applied to our standardized path coefficients. Here is what we get:

```
. estat stdize: test _b[math21:math7] == _b[math21:read7]
 ( 1)  [math21]math7 - [math21]read7 = 0
           chi2(  1) =     0.49
         Prob > chi2 =     0.4852
```

A chi-squared test of equality of the standardized coefficients yields a $\chi^2(1) = 0.49$, $p = 0.49$. You can see why this test is important. Although the effect of `math7` on `math21` in our sample was stronger, $\beta = 0.30$, than the effect of `read7` on `math21`, $\beta = 0.25$, this difference is not statistically significant. As with any test of significance, we are not saying that reading and math at age 7 are equally important, just that we have not demonstrated a significant difference using our data.

2.7   A cross-lagged panel design

Suppose we want to know the influence of math scores on reading skills and the influence of reading skills on math scores. One way to do this is to use panel data where we have at least two waves of data. We can use our example of math and reading achievement to assess this influence because we have data on both math and reading skills at age 7 and again at age 21. Our model is presented in figure 2.7.

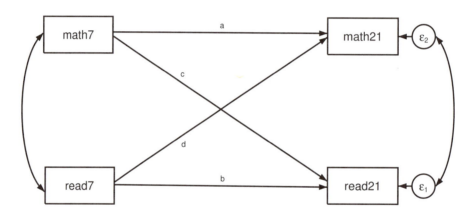

Figure 2.7. Cross-lagged panel design

We have a curved line between reading and math at age 7 because we assume these exogenous variables will be correlated.[7] Paths **a** and **b** are known as stability coefficients. They reflect how stable the corresponding concepts, math and reading, are over time. Given the huge age difference between ages 7 and 21, it is not obvious that these stability coefficients should be very large. The paths labeled **c** and **d** are of special interest: they tell us how much early math skills influence reading at age 21 and how much early reading skills influence math at age 21, respectively. Does a 7-year-old with good math skills benefit with better reading at age 21? Does a 7-year-old with good reading skills benefit with better math skills at age 21? Which path is stronger?

---

7. It is not necessary to draw these correlations when working with the SEM Builder; it could make the drawing quite messy if we had several exogenous variables. Remember, Stata assumes that these exogenous variables are correlated.

We must have correlated errors for `read21` and `math21` in our model. Some variance in `math21` that is not accounted for by `math7` and `read7` will surely be correlated with some variance in `read21` that is not accounted for by `math7` and `read7`. Some researchers choose not to include this correlation; however, it should be included, and the `sem` command makes it simple to estimate this correlation. Without this correlation, you are unlikely to obtain a good fit for your model, and excluding it would be equivalent to saying that all factors influencing `math21` are uncorrelated with those influencing `read21` except for `read7` and `math7`.

Here are the `sem` commands:

```
. sem (math21 <- math7 read7)          /// Eq. for math21
      (read21 <- read7 math7),          /// Eq. for read21
      method(mlmv) standardized          /// Missing, standardized
      cov(e.math21*e.read21)             // Covariances
  (output omitted)
Structural equation model                  Number of obs      =        416
Estimation method  = mlmv
Log likelihood     = -4369.2178
```

Standardized	Coef.	OIM Std. Err.	z	P>|z|	[95% Conf. Interval]	
Structural						
math21 <-						
math7	.3103359	.0469761	6.61	0.000	.2182645	.4024074
read7	.2571059	.0473963	5.42	0.000	.1642109	.3500008
_cons	1.960534	.2806437	6.99	0.000	1.410482	2.510585
read21 <-						
math7	.1102662	.0460747	2.39	0.017	.0199614	.200571
read7	.4931898	.0411844	11.98	0.000	.4124698	.5739098
_cons	6.343015	.4020612	15.78	0.000	5.55499	7.131041
mean(math7)	3.887371	.146786	26.48	0.000	3.599676	4.175067
mean(read7)	3.919858	.1489989	26.31	0.000	3.627825	4.21189
var(e.math21)	.7941389	.0379745			.7230916	.872167
var(e.read21)	.7149913	.0404764			.6399019	.7988922
var(math7)	1	.			.	.
var(read7)	1	.			.	.
cov(e.math21, e.read21)	.1735063	.0515344	3.37	0.001	.0725007	.2745119
cov(math7, read7)	.2722754	.0464696	5.86	0.000	.1811967	.3633541

```
LR test of model vs. saturated: chi2(0)   =       0.00, Prob > chi2 =     .
```

```
. estat eqgof
Equation-level goodness of fit
```

| | | Variance | | | | |
depvars	fitted	predicted	residual	R-squared	mc	mc2
observed						
math21	7.173452	1.476735	5.696717	.2058611	.4537192	.2058611
read21	71.36437	20.33946	51.02491	.2850087	.533862	.2850087
overall				.3767532		

```
mc  = correlation between depvar and its prediction
mc2 = mc^2 is the Bentler-Raykov squared multiple correlation coefficient
```

There are 0 degrees of freedom, so we have no test of significance for the model. Because of this, there are no modification indices to be reported. We also did not ask for indirect effects because we do not have any in the model. The results are summarized in figure 2.8.

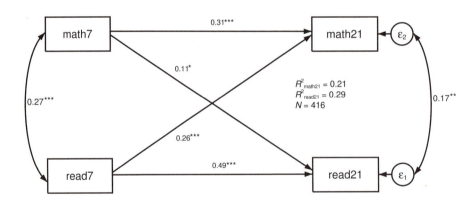

Figure 2.8. Estimated model for cross-lagged panel relating reading and math skills ($*p < 0.05$, $**p < 0.01$, and $***p < 0.001$)

What can we say about these results? First, it appears that reading skills are more stable than math skills ($\beta = 0.49$ compared with $\beta = 0.31$, both $p < 0.001$), although we should test whether there is a significant difference. Second, it appears that reading skills at age 7 influence math skills at age 21 ($\beta = 0.26$, $p < 0.001$) more so than math skills at age 7 influence reading skills at age 21 ($\beta = 0.11$, $p < 0.05$). We should also test whether these differences are statistically significant.

To test for significant differences between standardized path coefficients, we first run
`sem, coeflegend` to retrieve the names Stata has given the paths. Using these names,
we then run our tests. We precede each test with the prefix command `estat stdize:`
so that Stata runs the test on the standardized estimates. Here are the postestimation
commands and their results:

```
. sem, coeflegend
. estat stdize: test _b[math21: math7]  = _b[read21: read7]
 ( 1)  [math21]math7 - [read21]read7 = 0
           chi2(  1) =     7.98
         Prob > chi2 =     0.0047
. estat stdize: test _b[math21: read7]  = _b[read21: math7]
 ( 1)  [math21]read7 - [read21]math7 = 0
           chi2(  1) =     4.60
         Prob > chi2 =     0.0319
```

Thus we can assert that the stability coefficient for reading skills is significantly
different from the stability coefficient for math skills, $\chi^2(1) = 7.98$, $p < 0.01$. We can
also assert that the effect of reading on later math skills differs significantly from the
effect of math on later reading skills, $\chi^2(1) = 4.60$, $p < 0.05$. The chi-squared test is a
two-tail test. If you had directional hypotheses and wanted a one-tail test, you could
cut the estimated p-values in half.

You can apply a panel model like this to a wide variety of situations. Do wives'
political views influence their husbands' political views more than the other way around?
You might measure their political views at three time points: one year before the
election, six months before the election, and at the election. In this case, you would
simply extend the model in figure 2.7 to have three waves. You would repeat the cross-
lagged paths between the second and third waves, and you would correlate the error
terms for the second and third waves.

2.8   Moderation

In this chapter, I have focused on meditational models, but `sem` can be used equally well
for models that involve moderation. Statisticians use the term "statistical interaction"
when they want to show that the effect of one variable x on another variable y is
moderated by the level of a third variable z. People who have a high score on z might
have a stronger or weaker relationship between x and y, in which case we say that z
moderates the relationship between x and y.

Consider the relationship between a person's score on intelligence and their academic
performance. The third variable here might be motivation. Students who have low levels
of motivation may have a very weak relationship between their intelligence and their
academic performance. By contrast, students who are highly motivated might have a
strong relationship between their level of intelligence and their academic performance;
high academic performance goes with high intelligence, but only when the person is
motivated. In this example, motivation would be a continuous moderator.

There are also applications where you might have a categorical moderator. For example, you might develop an intervention program X designed to reduce math anxiety Y among first-year graduate students. Depending on how you design the program, the results may vary by gender Z. If your program involves math training with examples that are more familiar to men than they are to women, then the relationship between X and Y might be stronger for men than it is for women. The categorical Z variable, gender, is said to moderate the relationship between X and Y. Discovering that gender was a moderator would have important implications on how you might want to revise your intervention.

Let us consider a very simple example of moderation using the `auto.dta` dataset that comes with Stata. Suppose we are interested in the relationship between vehicle weight x and miles per gallon (MPG) y. We think that the greater the weight of a car, the worse its MPG. We need to recode `weight` to measure weight in 1,000s of pounds. Here are our commands:

```
. sysuse auto, clear
. generate wgt1000s = weight/1000
. sem (mpg <- wgt1000s)
. estat eqgof
```

Our model is shown in figure 2.9, where we have measured weight in 1,000s of pounds. Gas mileage goes down by 6.01 MPG for each additional 1,000 pounds of weight. Weight alone explains 65% of the variance in MPG.

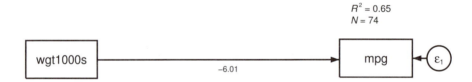

Figure 2.9. Bivariate relationship between MPG and weight in 1,000s

Next we think that foreign brand cars may get better mileage than domestic brand cars. We could incorporate both weight and foreign brand as joint predictors of MPG as illustrated in figure 2.10 by using

```
. sem (mpg <- wgt1000s foreign)
. estat eqgof
```

The addition of whether the brand is foreign does not help significantly once we have adjusted for weight. Thus it appears that the key variable for MPG is simply how much a car weighs.

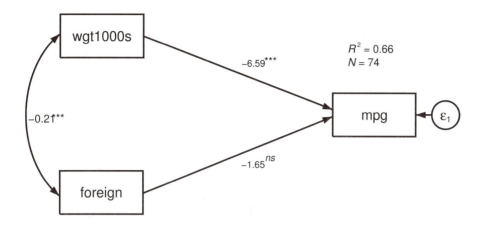

Figure 2.10. Weight and foreign brand as predictors of MPG (***$p < 0.001$; ns = not significant)

What about moderation? A researcher might ask whether foreign brand cars are designed with MPG as a primary goal. Foreign cars may also have differences that mitigate the effect of weight on MPG, for example, perhaps they have a different transmission or gear ratio. If this were the case, then the relationship between weight and MPG might be different for foreign brand cars than it is for U.S. brand cars. We predict that the relationship between weight and MPG will be weaker for foreign brand cars than for domestic brand cars. We need to generate an interaction term between `wgt1000s` and `foreign`, and then run the structural equation model including this interaction term:

```
. generate wgt1000sXforeign = wgt1000s*foreign
. sem (mpg <- wgt1000s foreign wgt1000sXforeign)
  (output omitted)
Structural equation model                   Number of obs      =         74
Estimation method   = ml
Log likelihood      = -298.90093
```

		OIM				
	Coef.	Std. Err.	z	P>|z|	[95% Conf.	Interval]
Structural						
mpg <-						
wgt1000s	-5.975083	.644035	-9.28	0.000	-7.237369	-4.712798
foreign	9.271333	4.377086	2.12	0.034	.6924014	17.85026
wgt1000sX~n	-4.450874	1.735688	-2.56	0.010	-7.85276	-1.048988
_cons	39.64696	2.18189	18.17	0.000	35.37054	43.92339
var(e.mpg)	10.22853	1.68156			7.410995	14.11724

```
LR test of model vs. saturated: chi2(0)   =      0.00, Prob > chi2 =     .
```

```
. estat eqgof
Equation-level goodness of fit
```

		Variance				
depvars	fitted	predicted	residual	R-squared	mc	mc2
observed						
mpg	33.01972	22.79119	10.22853	.6902297	.8308006	.6902297

(*output omitted*)

The intercept _cons = 39.65 is a bit difficult to interpret because this is the value when all the independent variables are at 0; that is, this would be a domestic car that weighed nothing. If we wanted a useful intercept, we might center weight by subtracting the mean weight from each car. This way, the new intercept would be the MPG for a domestic car that was of average weight for all cars.

Let us focus on the effects of wgt1000s, foreign, and wgt1000sXforeign, which is their interaction. Because our moderator is whether the car is foreign, we might think of the resulting equation for domestic cars and then for foreign cars:

Domestic cars

$$mpg\_domestic = 39.65 + 9.27(0) - 5.98(\texttt{wgt1000s}) - 4.45(\texttt{wgt1000s} \times 0)$$
$$= 39.65 - 5.98(\texttt{wgt1000s})$$

Foreign cars

$$mpg\_foreign = 39.65 + 9.27(1) - 5.98(\texttt{wgt1000s}) - 4.45(\texttt{wgt1000s} \times 1)$$
$$= 39.65 + 9.27 - 10.43(\texttt{wgt1000s})$$
$$= 48.92 - 10.43(\texttt{wgt1000s})$$

Thus a domestic car loses 5.98 MPG for each additional 1,000 pounds of weight. Surprisingly, a foreign car loses 10.43 MPG for each additional 1,000 pounds of weight. Controlling for weight, lighter foreign cars do get better mileage, but this advantage disappears for heavier foreign cars.

We can illustrate this in two ways. Figure 2.11[8] shows one way, where the car being foreign moderates the effect of weight on mileage.

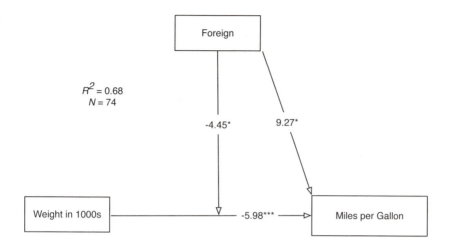

Figure 2.11. The relationship between weight and MPG as mediated by whether the car is a foreign brand ($*p < 0.05$ and $***p < 0.001$)

Perhaps a better way to illustrate how a foreign brand moderates the effect of weight on MPG is to use the **predict** command following **sem** and then generate a two-way graph. We simply run **predict mpgpf** (or a name of your choosing) to obtain a predicted value for **mpg** for each car, whether domestic or foreign, based on its weight and the interaction of these two predictors. The **twoway** graph command is long; you might want to use the dialog box to get it. We ask for a **scatterplot** (**scatter**) between **mpgpf** and **wgt1000s** separately for foreign cars, **foreign == 1**, and domestic cars, **foreign == 0**. In the rest of the command, we pick a color and define the symbol to represent foreign and domestic cars. Finally, we modify the default labels for the y axis, the x axis, and the legend.

```
. sem (mpg <- wgt1000s foreign wgt1000sXforeign)

. estat eqgof

. predict mpgpf

. twoway (scatter mpgpf wgt1000s if foreign==1, mcolor(black)
> msymbol(diamond_hollow))
> (scatter mpgpf wgt1000s if foreign==0, mcolor(black) msymbol(circle_hollow)),
> ytitle(Estimated Miles Per Gallon) xtitle(Car´s Weight in 1000s of pounds)
> legend(order(1 "Foreign" 2 "Domestic"))
```

---

8. This figure was not drawn using the SEM Builder because this freehand model was a better way to represent the data.

Our scatterplot, shown in figure 2.12, does a good job of showing the moderation effect. Foreign cars, represented by hollow diamonds, tend to weigh less than domestic cars; none of the foreign cars weigh more than about 3,500 pounds, while some domestic cars weigh nearly 5,000 pounds. As the weight goes up, the foreign cars have a sharper drop-off in their estimated MPG. Foreign cars, on average, get better gas mileage because most of them are relatively light, around 2,500 pounds, whereas most U.S. cars are over 3,000 pounds.

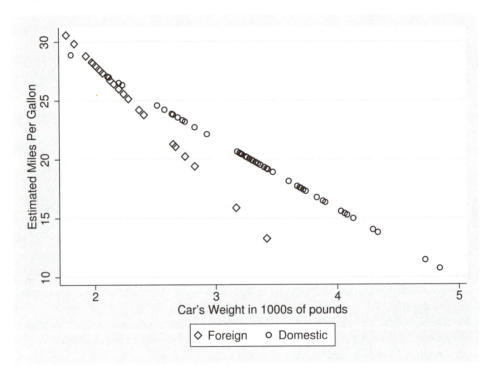

Figure 2.12. A scatterplot of predicted values between weight and MPG as mediated by whether the car is a foreign brand

We could do all this with the `regress` command, but using the `sem` command has a couple of advantages. First, if we had missing values, we could use the `method(mlmv)` option to automatically handle missing values better than we would by using listwise deletion with the ordinary least-squares command if it were reasonable to assume multivariate normality. Second, after you complete chapter 3 using the full structural equation modeling design involving both a measurement model and a structural model, we can take advantage of the ability to estimate measurement errors and have correlated measurement errors where appropriate. Finally, we can extend the structural equation model to include much more complex moderation models. When the moderator is categorical, we sometimes treat the categories as separate groups and do a multigroup analysis, as you will learn in chapter 5.

Here are just a couple of examples of these extensions. First, the moderator may interact with several exogenous variables, and we might have several outcomes. Suppose we are using prestige of undergraduate school, scores on the GRE, and undergraduate grade point average to predict a graduate student's years to completion of a PhD. We might consider motivation to be a continuous moderator and marital status to be a categorical moderator. Even though this is a more complex model, we could still fit it by using ordinary least-squares regression.

Now suppose we are predicting academic performance of graduate students, and we have three indicators: grades, time to complete degree, and number of papers presented at professional meetings. Suppose our moderator is the expectation of the student's significant others, meaning his or her peers, parents, teachers, etc. To deal with the moderation model, we need to wait until we complete chapter 3, but the bottom line is that we could not do it with ordinary least-squares regression.

We might also want to fit what is known as a moderated mediation model. In such a model, the indirect effects mediating between x and y will vary with the moderating variable. For women, the indirect effects may be stronger than the direct effects. For men, the direct effects might be stronger than the indirect effects.

2.9 Nonrecursive models

So far, we have focused on recursive models where there is neither feedback nor reciprocal paths. Sometimes, there is reciprocal influence between variables. Whenever you have dyadic data, reciprocal paths are likely. We can imagine a dyad where the influence is strictly hierarchical, perhaps in a military style of organization. More likely, influence goes both down and up. In interpersonal relationships, there may not even be a down or up hierarchy and the influence is truly mutual.

Whenever there is a reciprocal relationship or some other form of feedback, we say the model is nonrecursive. Consider the marital satisfaction of wives and their husbands. We believe that the more satisfied the wife is, the more satisfied the husband will be, her_sat → his_sat. At the same time, we believe that the more satisfied the husband is, the more satisfied the wife will be, his_sat → her_sat. We represent this nonrecursive relationship in figure 2.13.

By examining figure 2.13, we can see that we have far too little information to fit the model. We have too many parameters to estimate with just a pair of observed variables. Can you explain why we have correlated error terms in the figure? Remember that ϵ_1 represents the unexplained variance in the wife's satisfaction and ϵ_2 represents the unexplained variance in the husband's satisfaction. Couples share so many variables in common that will be influencing their mutual satisfaction; they face the same economic challenges, neighbor problems, parenting issues. You can generate a long list of factors that influence the satisfaction of both a wife and her husband. None of these are included in our model, so we must at least acknowledge this by allowing the errors to be correlated.

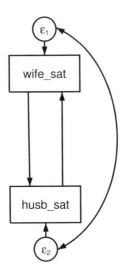

Figure 2.13. Reciprocal relationship between a wife's satisfaction and her husband's satisfaction: A nonrecursive model (unidentified)

There is a technical problem with this nonrecursive model. We cannot use `wife_sat` by itself to predict `husb_sat`, because the wife's satisfaction depends on the husband's, and vice versa. If she is highly satisfied, this leads to him being more satisfied, which leads in turn to her being more satisfied. We will see how this creates a potential instability in the model.

One way to fit this model is to use what are known as instrumental variables. These variables directly influence the wife's satisfaction but do not directly influence her husband's satisfaction, and vice versa. We use the instrumental variables to predict the wife's satisfaction in a way that does not influence her husband's satisfaction directly. We can then use this predicted value of her satisfaction to estimate the effect of her satisfaction on his satisfaction. We do the same for the husband. The problem with this approach is that it can be very difficult to locate variables that directly influence one but not both of our reciprocally related variables.

The instrumental variable could have an indirect effect. For example, you might use the husband's education as the instrumental variable for `husb_sat`. His own education might predict his satisfaction but not directly influence his wife's satisfaction. If he has more education, he is likely to make more income, which may indirectly make the wife more satisfied. Then we could use the wife's education as the instrumental variable for her satisfaction, making the same argument in reverse.[9] Our model might look like figure 2.14.

---

9. This model will be limited by how compelling this argument is. Certainly, there are many couples who are satisfied where neither wife nor husband have much education, and there are also many couples who are miserable though wife and husband both have substantial education.

Although this model is identified, it is not without problems. The quality of the solution depends on finding good instrumental variables. Unless you have a very large sample, you want the instrumental variables to have fairly strong relationships with the reciprocally related variables. The correlations between wife_ed and wife_sat or husb_ed and husb_sat are probably modest. Equally important, the correlation between wife_ed and husb_ed is probably very high. This creates a collinearity problem. If we had data on a modest-size sample to fit this model, the standard errors on the reciprocal paths would be quite large, reducing our confidence in the estimated parameters. You can mitigate the problem of having a weak instrumental variable by having a very large sample. Otherwise, even if you get fairly substantial estimates for the reciprocal effects, they likely will not be statistically significant.

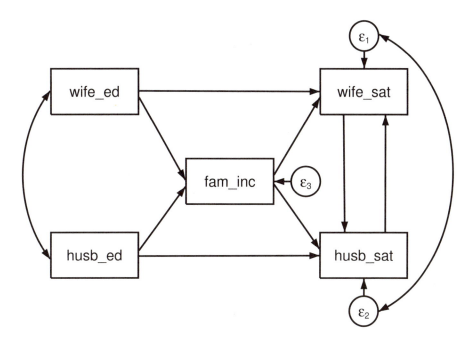

Figure 2.14. Reciprocal relationship between a wife's satisfaction and her husband's satisfaction: A nonrecursive model (identified)

2.9.1 Worked example of a nonrecursive model

We will use example 7 from Stata's *Structural Equation Modeling Reference Manual*. This example appeared originally in Duncan, Haller, and Portes (1968). The purpose of this study was to try to estimate the reciprocal influence that friends have on one another. A sample of 329 youth reported their occupational aspirations. Their friends were then interviewed and also reported their own occupational aspirations. With just these two variables, we have the problem we illustrated in figure 2.13 of too many parameters and too little information to estimate them. The researchers also obtained a measure of the respondent and the friend on intelligence and socioeconomic status. They felt that the respondent's socioeconomic status would certainly influence his or her aspirations but might also influence the friend's aspirations, and vice versa. However, they felt that the respondent's intelligence would only directly influence the respondent's own aspirations, while the friend's intelligence would only directly influence the friend's aspirations. Their model appears in figure 2.15.

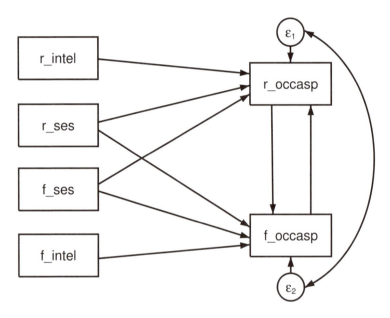

Figure 2.15. Reciprocal relationship between respondent and friend for occupational aspirations

Do you like this model? Do you agree with the researchers' assumptions? Pretend you are the respondent. What they are saying, for example, is that your friend's intelligence only indirectly influences your aspirations by first directly influencing your friend's aspirations. If your best friend Susan is extremely intelligent, she is likely to aspire to a fairly high occupational status. As she shares these aspirations with you, this may expand your horizons about your own possible occupational status, regardless of how intelligent you are and regardless of your socioeconomic status.

This nonrecursive model is identified because we have a clear instrumental variable, r_intel, that directly influences your own aspirations, r_occasp, but does not directly influence your friend's aspirations, f_occasp. We also have a clear instrumental variable, f_intel, that directly influences your friend's aspirations, f_occasp, but does not directly influence your own aspirations, r_occasp.

We can use the SEM Builder or the sem command directly to fit this model. This command specifies each path in the model. The covariance between the errors appears as an option, and we ask for a standardized solution.

```
. use http://www.stata-press.com/data/r13/sem_sm1.dta, clear
(Structural model with all observed values)

. sem (r_occasp <- f_occasp r_intel r_ses f_ses)
>     (f_occasp <- r_occasp f_intel f_ses r_ses),
>     cov(e.r_occasp*e.f_occasp) standardized
```
(output omitted)

Standardized	Coef.	OIM Std. Err.	z	P>|z|	[95% Conf. Interval]	
Structural						
r_occasp <-						
f_occasp	.2773441	.1281904	2.16	0.031	.0260956	.5285926
r_intel	.2854766	.05	5.71	0.000	.1874783	.3834748
r_ses	.1570082	.0520841	3.01	0.003	.0549252	.2590912
f_ses	.0973327	.060153	1.62	0.106	-.020565	.2152304
f_occasp <-						
r_occasp	.2118102	.156297	1.36	0.175	-.0945264	.5181467
r_ses	.0794194	.0587732	1.35	0.177	-.0357739	.1946127
f_ses	.1681772	.0537199	3.13	0.002	.062888	.2734663
f_intel	.3693682	.0525924	7.02	0.000	.2662891	.4724474
var(e.r_occ~p)	.6889244	.0399973			.6148268	.7719519
var(e.f_occ~p)	.6378539	.039965			.5641425	.7211964
cov(e.r_occ~p, e.f_occasp)	-.2325666	.2180087	-1.07	0.286	-.6598558	.1947227

LR test of model vs. saturated: chi2(0)   =   0.00, Prob > chi2 =   .

Figure 2.16 shows the results for the SEM Builder. When we used the SEM Builder, we did not draw the correlations between the exogenous variables, but Stata assumes these observed exogenous variables are correlated. We can draw them using the SEM Builder, but the results we get will be the same. If you publish a figure like this, you typically add a footnote indicating that the correlations among the independent variables are not shown to simplify the presentation.

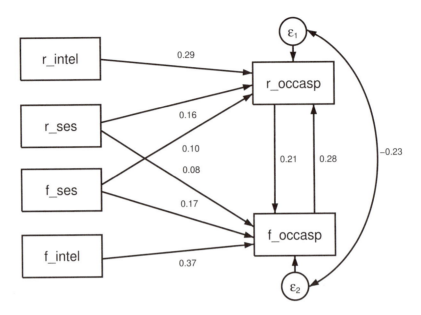

Figure 2.16. Estimated reciprocal relationship between respondent and friend for occupational aspirations ($N = 329$)

These results show a modest reciprocal influence. For f_occasp $\rightarrow$ r_occasp, the $\beta = 0.28$; $z = 2.15$, $p < 0.05$. Going the other way, the $\beta = 0.21$; $z = 1.35$, $p = 0.18$, so this path is not significant, even if it is a fairly similar value. Again, it can be difficult to obtain significant results for reciprocal paths unless you have strong instrumental variables. In the next section, we will apply some interesting constraints on the model, and the results will be more interesting.

2.9.2   Stability of a nonrecursive model

Previously, I indicated that there is a potential technical problem with a reciprocal linkage in that it may not be stable. Our results here are standardized (we have no choice because we only had a correlation matrix to use in our analysis). Imagine we had the actual raw data and fit the model for an unstandardized solution. Suppose the path f_occasp $\rightarrow$ r_occasp had $\beta = 1.2$, and the path r_occasp $\rightarrow$ f_occasp had $\beta = 1.5$. If your aspirations go up 1 unit, your friend's go up 1.2 units. This makes your aspirations go up $1.2 \times 1.5 = 1.8$ units, which causes your friend's aspirations to go up $1.8 \times 1.2 = 2.16$ units, which causes your aspirations to go up $2.16 \times 1.5 = 3.24$ units, and so on. These simultaneous changes will lead to an explosive result. When this happens, we say the model is unstable.

Alternatively, imagine you had unstandardized values of 0.2 and 0.3 for the pair of β's. Then when your aspirations go up 1 unit, your friend's go up 0.2 units. This makes your aspirations go up $0.2 \times 0.3 = 0.06$ units, which makes your friend's aspirations go up a further $0.06 \times 0.2 = 0.012$ units, and so on. In this case, the model will stabilize as the subsequent increments become smaller and eventually converge on 0.000.

When you have unstandardized estimates, you can run a command to verify the stability of your model. In our example, the command `estat stable` produces a stability index. If this index is less than 1.0, you have a stable model. All recursive models are stable, but nonrecursive models may or may not be stable.

2.9.3 Model constraints

Take a close look at our model. Is there any reason why a randomly selected respondent should have more influence on a friend than a friend has on the respondent? Because our sample is treated as a random sample, there is no reason why these reciprocal paths should not be identical. The same case can be made for the effects of each of the exogenous variables. Certainly there will be some variation in a particular sample like this, but logically the population values should be equal. A randomly selected respondent's intelligence should have no more effect on that respondent's aspirations than a friend's intelligence has on the friend's aspirations. This also applies to the effects of socioeconomic status. We expect the following set of equalities to hold:

$$
\begin{aligned}
\texttt{r\_intel -> r\_occasp} &= \texttt{f\_intel -> f\_occasp} \\
\texttt{r\_ses -> r\_occasp} &= \texttt{f\_ses -> f\_occasp} \\
\texttt{f\_ses -> r\_occasp} &= \texttt{r\_ses -> f\_occasp} \\
\texttt{f\_occasp -> r\_occasp} &= \texttt{r\_occasp -> f\_occasp}
\end{aligned}
$$

When imposing equality constraints, we need to switch from a standardized solution to an unstandardized solution. If the unstandardized solutions have equal estimates for a parameter, then the structural relationship is the same. When this happens, the standardized estimates will also be equal only when the corresponding variances are equal.

To make these four equality tests, we need to first know the nicknames Stata assigns to each of the eight paths. We can find these labels by running `sem, coeflegend`. Here are partial results:

```
. sem, coeflegend

Structural equation model                    Number of obs     =      329
Estimation method  = ml
Log likelihood     = -2617.0489
```

	Coef.	Legend
Structural		
r_occasp <-		
f_occasp	.2773441	_b[r_occasp:f_occasp]
r_intel	.2854766	_b[r_occasp:r_intel]
r_ses	.1570082	_b[r_occasp:r_ses]
f_ses	.0973327	_b[r_occasp:f_ses]
f_occasp <-		
r_occasp	.2118102	_b[f_occasp:r_occasp]
r_ses	.0794194	_b[f_occasp:r_ses]
f_ses	.1681772	_b[f_occasp:f_ses]
f_intel	.3693682	_b[f_occasp:f_intel]
var(e.r_occ~p)	.6868304	_b[var(e.r_occasp):_cons]
var(e.f_occ~p)	.6359151	_b[var(e.f_occasp):_cons]
cov(e.r_occ~p, e.f_occasp)	-.1536992	_b[cov(e.r_occasp,e.f_occasp):_cons]

```
LR test of model vs. saturated: chi2(0)    =      0.00, Prob > chi2 =      .
```

Suppose we are only interested in testing the reciprocal linkage between `r_occasp` and `f_occasp`. The output above reports the unstandardized values: 0.28 and 0.21. Because we are using an unstandardized solution, we do not need the prefix command `estat stdize:`. Our test is simply as follows:

```
. test (_b[r_occasp:r_intel] == _b[f_occasp:f_intel])
 ( 1)  [r_occasp]r_intel - [f_occasp]f_intel = 0
          chi2( 1) =     1.43
        Prob > chi2 =     0.2323
```

Thus even though one path was significant and the other was not, the two unstandardized estimates are not significantly different from each other: $\chi^2(1) = 1.43$, $p = 0.23$.

We can test all four equalities simultaneously by simply extending our command to include the labels for the other paths:

```
. test (_b[r_occasp:r_intel ]==_b[f_occasp:f_intel ])
> (_b[r_occasp:r_ses ]==_b[f_occasp:f_ses ])
> (_b[r_occasp:f_ses ]==_b[f_occasp:r_ses ])
> (_b[r_occasp:f_occasp]==_b[f_occasp:r_occasp])
 ( 1)   [r_occasp]r_intel - [f_occasp]f_intel = 0
 ( 2)   [r_occasp]r_ses - [f_occasp]f_ses = 0
 ( 3)   [r_occasp]f_ses - [f_occasp]r_ses = 0
 ( 4)   [r_occasp]f_occasp - [f_occasp]r_occasp = 0
            chi2(  4) =     1.61
        Prob > chi2 =     0.8062
```

Based on these results—$\chi^2(4) = 1.61$, $p = 0.81$—we can conclude that the corresponding paths are not significantly different whether we are talking about effects of respondents on their friends or vice versa.

2.9.4 Equality constraints

While these results tell us that corresponding paths are not significantly different, the results do not estimate the shared value each pair of paths has. Stata can constrain specific paths to be equal by attaching the same name to each of them. The name can be from 1 to 32 characters long. For example, to constrain r_occasp $\rightarrow$ f_occasp to be the same value as f_occasp $\rightarrow$ r_occasp, we could use the label b1. Here is how we change our command:[10]

---

10. This command would normally be applied to the unstandardized solution, but the data we are using for this example are limited to a correlation matrix. Nonetheless, we drop the standardized option. When we leave this option, we get slight differences on one of the pairs that we constrained to be equal. This is because the standardization uses the fitted variances, and they are not exactly equal to 1.0.

```
. sem (r_occasp <- f_occasp@b1 r_intel@b2 r_ses@b3 f_ses@b4)
>      (f_occasp <- r_occasp@b1 f_intel@b2 f_ses@b3 r_ses@b4),
>      cov(e.r_occasp*e.f_occasp)
 (output omitted)
```

```
Structural equation model                    Number of obs      =         329
Estimation method  = ml
Log likelihood     = -2617.8705
 ( 1)  [r_occasp]f_occasp - [f_occasp]r_occasp = 0
 ( 2)  [r_occasp]r_intel - [f_occasp]f_intel = 0
 ( 3)  [r_occasp]r_ses - [f_occasp]f_ses = 0
 ( 4)  [r_occasp]f_ses - [f_occasp]r_ses = 0
```

		OIM				
	Coef.	Std. Err.	z	P>|z|	[95% Conf.	Interval]
Structural						
r_occasp <-						
f_occasp	.2471578	.1024504	2.41	0.016	.0463588	.4479568
r_intel	.3271847	.0407973	8.02	0.000	.2472234	.4071459
r_ses	.1635056	.0380582	4.30	0.000	.0889129	.2380984
f_ses	.088364	.0427106	2.07	0.039	.0046529	.1720752
f_occasp <-						
r_occasp	.2471578	.1024504	2.41	0.016	.0463588	.4479568
r_ses	.088364	.0427106	2.07	0.039	.0046529	.1720752
f_ses	.1635056	.0380582	4.30	0.000	.0889129	.2380984
f_intel	.3271847	.0407973	8.02	0.000	.2472234	.4071459
var(e.r_occ~p)	.6884513	.0538641			.5905757	.8025477
var(e.f_occ~p)	.6364713	.0496867			.5461715	.7417005
cov(e.r_occ~p,						
e.f_occasp)	-.1582175	.1410111	-1.12	0.262	-.4345942	.1181592

```
LR test of model vs. saturated: chi2(4)   =      1.64, Prob > chi2 = 0.8010
. estat eqgof
Equation-level goodness of fit
```

		Variance				
depvars	fitted	predicted	residual	R-squared	mc	mc2
observed						
r_occasp	1.020027	.3252899	.6884513	.3250653	.5701711	.3250951
f_occasp	.9756657	.3396728	.6364713	.3476543	.5896223	.3476545
overall				.4952231		

```
mc  = correlation between depvar and its prediction
mc2 = mc^2 is the Bentler-Raykov squared multiple correlation coefficient
```

These equality constraints are entirely appropriate and simplify our model considerably. We now have 4 degrees of freedom instead of 0: $\chi^2(4) = 1.64$, $p = 0.80$. The reciprocal effects have equal values because of our equality constraint on them. Importantly, both `r_occasp` $\rightarrow$ `f_occasp` and `f_occasp` $\rightarrow$ `r_occasp` are statistically significant: $z = 2.41$, $p < 0.05$.

When you are analyzing raw data, there is likely to be a difference between the standardized values of the parameter estimates even when the unstandardized values are constrained to be equal. This is because the standardized values depend both on the structure of the relationship (the unstandardized value) and the relative variance of the variables. If some variables have different variances, then the standardized values will be unequal. Because we are comparing the structural values for the two sets of coefficients, we can ignore the differences in the standardized values.

The postestimation commands you would also want to run on this model should be familiar by now. We did include `estat eqgof` to obtain the R^2 values, but you should also run `estat teffects, standardized` to obtain the indirect and total effects, and `estat mindices` to obtain the modification indices. You may also want to run `estat eqgof` with the `mc2` coefficients as your estimates of explained variance instead of R^2 (Bentler and Raykov 2000). This is a variation of R^2 designed exclusively for nonrecursive models. Like other correlations, the `mc` coefficient can be negative. Because the square of this must be positive, you should check whether the `mc` value is negative.

2.10 Exercises

1. Draw a path model that includes six variables with at least one variable that is an endogenous mediator. Use variables from your substantive area of interest and make sure you are able to defend your path model.

   a. Identify which variables are exogenous.

   b. Identify which variables are endogenous mediators.

   c. Identify which variables are endogenous outcomes.

   d. Why do we not use the traditional labels of a variable being either independent or dependent?

2. Label the paths in your model as shown in figure 2.1.

   a. List each direct effect.

   b. List each possible indirect effect, and show how it is a product of direct effects.

3. The traditional approach to mediation assumes that if x is uncorrelated with y, then there is no mediation.

   a. Draw a model in which the correlation of x and y might be 0, but there is still mediation.

   b. Explain how there can be mediation when there is no bivariate correlation.

4. What command gives you indirect effects and total effects?

5. What command gives you the R^2 for each endogenous variable?

6. Why would you correlate error terms for endogenous mediator variables?

7. You have a large company that has a serious problem with employee absenteeism. You are considering a company-wide, employer-sponsored exercise program. You feel that exercise directly reduces stress, improves overall health, and decreases absenteeism. You feel that stress directly reduces health and increases absenteeism. Finally, you feel that improved health reduces absenteeism. The following table shows hypothetical correlations and standard deviations. These are based on a pilot test of 200 employees.

	exercise	age	stress	health	absenteeism
exercise	1.0				
age	−0.2	1.0			
stress	−0.4	0.1	1.0		
health	−0.5	−0.2	0.3	1.0	
absenteeism	−0.3	0.1	0.4	0.5	1.0
sd	3.1	3.4	3.8	2.9	3.0

   a. Draw this as a path model by using the SEM Builder. Fit the model by using commands (not the SEM Builder).

   b. Estimate a standardized solution to the model.

   c. How well does your model fit the data?

   d. How much variance (R^2) do you explain for each endogenous variable?

   e. What is the direct, indirect, and total effect of exercise on absenteeism? How significant is each?

   f. Provide a summary of your results, including a table that you could show the Board of Directors to convince them to sponsor a company-wide exercise program.

8. Draw the model in exercise 7, and fit the result by using the SEM Builder.

   a. Is there a significant difference between the direct effect of health and the direct effect of stress on absenteeism?

   b. Compare the results in the model you fit above with the results from a model that only includes the direct effects of exercise, stress, and health on absenteeism. Why is a model that includes only direct effects misleading? What does the total effect tell you that you would miss if you did a simple regression of absenteeism on the three predictors?

9. Name two variables that are reciprocally related and explain why they are reciprocally related. Name some instrumental variables that would allow you to estimate the reciprocal relationship. Do a freehand drawing of the model, and justify why the instrumental variables are appropriate.

2.A   Using the SEM Builder to run path models

To get started, let us first open the dataset:[11]

```
. use http://www.stata-press.com/data/dsemusr/path.dta, clear
```

We will first draw and then fit the model that appears in figure 2.7. With the dataset open, type `sembuilder` into the Command window and press *Enter*.[12] If you have changed settings in your SEM Builder when constructing an earlier model, you may want to restore the default settings that control how the results will be displayed. Click on **Settings > Settings Defaults**.

We need to have four observed variables: `math7`, `read7`, `math21`, and `read21`. Click on the **Add Observed Variable** tool, □, and then click on the screen where you want each variable's rectangle to appear. Make sure to allow room for the error terms and their correlations on the right side and the correlation on the left side.

Click on the **Select** tool, ▸, in the upper left of the screen, and then click and drag over the top two rectangles to select them. Once these are highlighted, click on **Object > Align > Horizontal Center**. Do the same for the bottom two rectangles. You can also use **Object > Align** to vertically align the two rectangles on the left as well as the two rectangles on the right.

Using the **Select** tool, click on the rectangle at the top left. Use the **Variable** drop-down menu on the toolbar to choose `math7` as name of the variable for this rectangle. Repeat this process to specify the variable names for the other three rectangles.

Next click on the **Add Path** tool, ⟶, and draw the four paths. When you draw these, the SEM Builder automatically adds the error terms, ϵ_1 and ϵ_2, for the endogenous variables. Click on the **Add Covariance** tool, ⌢, and connect the two error terms. Stata will assume the exogenous variables, `math7` and `read7`, are correlated, so we do not need to draw the curve between them (but we will anyway for practice). When you use the icon for a curved line, the line will curve inward or outward depending on whether you connect going up or down. If it curves the wrong way, you can click on the **Select** tool, and then click on the curve. A blue line with a small, hollow circle at one end will appear on the curve. Click on the circle to change the curvature of the line.

---

11. You may also want to reread the chapter 1 appendix box 1.A, which details the differences between the SEM Builder interfaces in Stata for Windows, Unix, and Mac.
12. I like to make the SEM Builder fill most of my screen. You can do this by dragging the edge of the Builder. Then press the **Fit in Window** button on the toolbar to make the canvas fit the SEM Builder window.

You should have something similar to this now:

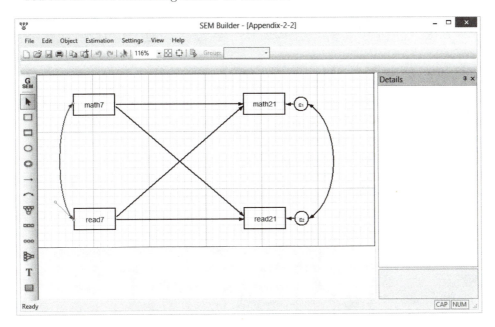

Now you are ready to fit the model. Click on **Estimation > Estimate....** On the **Model** tab select *Maximum likelihood with missing values*. Select the tab labeled **Reporting**, and check *Display standardized coefficients and values*. Then click on **OK** to produce the results in the Results window and put the standardized path coefficients on your figure (see figure 2.17). If you get the unstandardized solution because you did not select standardized coefficients, you can click on **View > Standardized Estimates** to get the standardized solution.

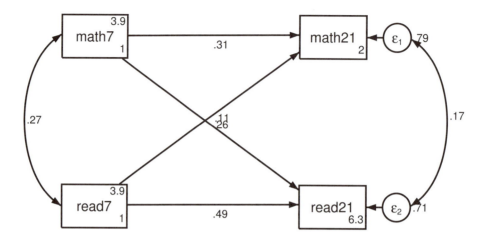

Figure 2.17. Model fit with standardized path coefficients displayed

This drawing has four features we would like to change:

1. A lot of extra information (intercepts, means, variances) is cluttering up the figure.

2. We would like all results to be rounded to two decimal places.

3. The paths for `read7` → `math21` and `math7` → `read21` have values located where it is hard to know which value goes with which path.

4. The significance levels and measures of goodness of fit are not displayed.

Let us deal with these problems in order. To reduce clutter, click on **Settings** > **Variables** > **All Observed...**,[13] and select the **Results** tab in the dialog box. In the *Exogenous variables* section, change both the `Variance` and the `Mean` options to `None`, and do the same for the `Intercept` option in the *Endogenous variables* section. Click on **OK**.

The model has a variance reported for each error term. My preference is to delete the error-variance numbers from the figure, keeping just the correlation between error terms wherever there is one. Go to **Settings** > **Variables** > **All Error...**, and on the **Results** tab change `Error variance` to `None`.

---

13. If you ever have variable names that are too long to fit in your rectangles, this is also where you fix that problem. Simply go to the **Box/Oval** tab, and change the size of the rectangles. Alternatively, you can click on the variable, go to the **Appearance** tab, check the box for *Customize appearance for selected variable*, click on **Set custom appearance**, and then adjust the size of the rectangle or oval.

To change the number of decimal places reported for the path coefficients and correlations, click on **Settings** > **Connections** > **All...**. In the dialog box that opens, go to the **Results** tab and click on **Result 1...**. Another dialog box will open, and here you can change the *Format* to %5.2f and then click on **OK**.

Moving the values for the paths is a bit tricky. So far, you have something that looks like the figure below. In the figure, I have used the **Select** tool to highlight the path from math7 to read21. This causes the path coefficient 0.11 to appear in blue and a button labeled **Properties...** to appear on the top toolbar.

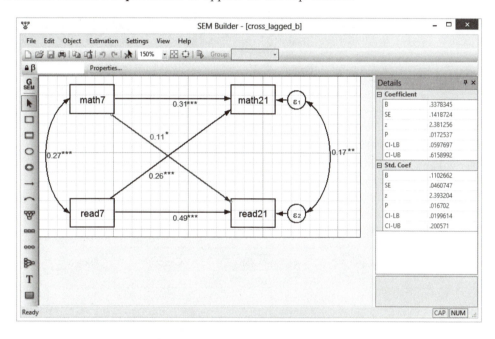

Click on the **Properties...** button, and go to the **Appearance** tab of the dialog box. Check *Customize appearance for selected connections*, and then click on **Set custom appearance** to open a new dialog box. Go to the **Results** tab, and click on the **Result 1...** button on the lower left side. One element of the dialog box that opens (see below) is *Distance between nodes*, which is a percentage of the distance between the predictor and its outcome. The default distance of 50% causes a problem in our case because this is where our paths cross each other. Change the percentage to 30% and click on **OK**.

Repeat the procedure for the path from `read7` to `math21` by first using the **Select** tool to highlight the path. Once you have changed the percentage for that path, we have taken care of our second problem.

To fix our final problem, we need to add textual information to our model. We will use one asterisk to denote a path that is significant at the 0.05 level, two asterisks to denote the 0.01 level, and three asterisks to denote the 0.001 level. Click on the **Add Text** tool on the left toolbar, and then click next to a coefficient. Insert the appropriate number of asterisks based on the p-value reported for the output in our results, which appear in the Results window of Stata. Do this for each coefficient. After entering the asterisks in their approximate locations, click on the **Select** tool, click on a set of asterisks, and drag it to exactly where you want it. Now you have a drawing that looks similar to figure 2.18.

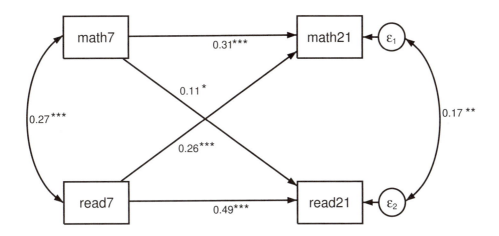

Figure 2.18. Model fit with less clutter and moved path values

Now we can enter the text for the model fit. Text can be included on the figure itself or in a footnote. You will learn how to include it in the figure in appendix A at the end of the book, but for now you may want to skip this part and just put the text in a footnote. In either case, we need to run `estat eqgof` as well as `estat gof, stats(all)`. We can run these using the SEM Builder. Click on **Estimation > Goodness of fit > Equation-level goodness of fit**. This gives us many options, but `Equation-level goodness-of-fit statistics (eqgof)` is already selected for us. Click on **OK**. Looking at the Results window, we see the R^2 values of 0.206 for `math21` and 0.285 for `read21`. Next click on **Estimation > Goodness of fit > Overall goodness of fit**. In the dialog box, use the drop-down menu under *Statistics to be displayed* to select `All of the above`. Click **OK**. If you also want modification indices, select **Estimation > Testing and CIs > Modification indices**.

3   Structural equation modeling

3.1   Introduction

In this chapter, we will learn about structural equation modeling (SEM), which is a combination of what you learned in chapter 1 about measurement models and what you learned in chapter 2 about path analysis and structural models. SEM opens up an enormous range of research capabilities. This chapter will complete your basic understanding of SEM, and then we will learn about some of the specific applications of SEM in the next chapters. SEM techniques can be applied to an ever-expanding range of research topics and offer highly flexible strategies.

The measurement model offers the ability to use multiple indicators of each latent variable and to isolate measurement error. Removing the measurement error from your latent variables allows stronger predictive power because measurement error is assumed to be a random error and as such has no explanatory power. The result is that the estimates of the path coefficients will usually be larger than if you had assumed no error in predictors, as we assume with traditional regression models. The path analysis part of the model is called the structural model, and it shows the theoretically causal linkages between the latent variables. The difference between this and the path analysis described in chapter 2 is simply that each observed variable in a path model can be a latent variable in full SEM. Remember that the causality in these models is in theory: the data can be consistent and hence support our causal theory, or the data can be inconsistent and lead us to modify our theory.

3.2   The classic example of a structural equation model

A classic structural equation model that has been used by most software packages as an example goes back to one of the earliest applications of SEM in the social sciences. Before SEM was available, many researchers argued that attributes such as attitudes and beliefs were too unstable to be useful. The observed correlation between a person's attitudes and beliefs at time A and time B was often so small that it called into question the value of studying these attributes. This frustration about the instability of attitudinal variables extends to many social–psychological concepts.

Consider the latent concept of alienation. This was seen as varying from day to day if not from hour to hour, and certainly from one year to the next. The argument was made that many social–psychological concepts were so unstable that we should ignore them. It makes little sense to predict a variable or to use it as a predictor if the concept that it represents is constantly changing.

Wheaton et al. (1977) developed a model of the stability of alienation over a 5-year period (1967–1971) and included socioeconomic status (SES) in 1966, SES66, as a covariate. They treated all three variables, alienation in 1967, alienation in 1971, and SES66 as latent variables. Anomia and powerlessness measured in 1967 and again in 1971 were the only two indicators of alienation, while education and occupational prestige served as indicators of SES66. When this was published in 1977, it was truly an innovative breakthrough.

Though it would have been better to have at least three indicators of alienation and preferably four, anomia and powerlessness are easily understood as reflective indicators of alienation, which is important for a latent variable. A person who has a high level on the latent variable (alienated) will therefore respond to items indicating that they are high on the observed variables (anomia and powerlessness). Both anomia and powerlessness depend on alienation, as shown in figure 3.1, and therefore are appropriate indicators. The direction of the arrows from each latent variable to its observed indicators is crucial.

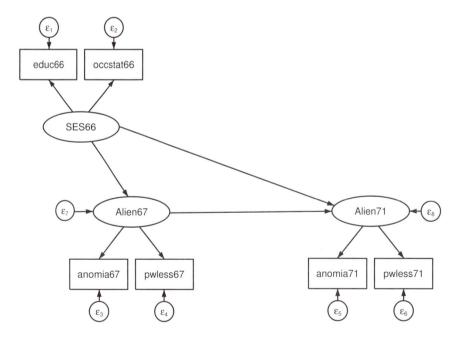

Figure 3.1. A structural equation model

The measurement model is more difficult to justify in the case of SES66. One could argue that in the case of SES66 the arrows are going in the wrong direction; that is, a person who has higher education and occupational status will therefore have higher SES66. A change in SES66 does not directly cause your education or occupational status; rather, your SES66 is a composite variable based on your education and occupational status. Consider a 50-year-old man who has very low SES66. Suppose he dropped out of high school. In this case, his low education (that ended 30+ years ago) contributed to his low SES66 (educ66 → SES66). We would not say that this man's SES66 explained his educational attainment 30 years ago.

The direction of the arrow is critical for understanding the meaning of a latent variable. The observed measures should reflect the latent variable rather than the other way around. We discuss this issue more later in this chapter. Another problem is that occupational status, in part, is measured by your level of education. This confounding of the two indicators is problematic. For now, let us take the model as Wheaton et al. (1977) presented it. We will appreciate their pioneering contribution to the diffusion of SEM throughout the social, behavioral, and health sciences.

This model can help us evaluate the question of whether alienation is extremely unstable or not. It may be that research before the Wheaton et al. article was published ignored measurement error. If we can improve how we handle measurement error (as opposed to blithely assuming our measurements are perfect), perhaps we can demonstrate reasonable stability of the core concept of alienation.

3.2.1 Identification of a full structural equation model

Figure 3.1 shows the initial 1977 model. In chapter 1, we discussed a preference of three indicators for a latent variable, but we also determined that you could use two if there were other indicators in the model. Let us examine our counting rule. We have six observed variables (indicators): anomia67, pwless67, anomia71, pwless71, educ66, and occstat66. Therefore, the variance–covariance matrix has $6(6+1)/2 = 21$ elements. The matrix contains 6 variances for our six indicators plus 15 covariances among our indicators. How many unknowns do we have? As we did in chapter 1, we arbitrarily fix a loading at 1.0 for the first indicator of each latent variable. The fixed loading is our reference indicator. This leaves us with just three loadings to estimate: Alien67 → pwless67, Alien71 → pwless71, and SES66 → occstat66. For our measurement model, we also have to estimate variances for six error terms, ϵ_1 to ϵ_6. The measurement model part of our structural equation model therefore has nine unknowns.

What about the structural part of our structural equation model? Alienation in 1967 and alienation in 1971 are endogenous latent variables because both of them depend on SES in 1966. Each endogenous latent variable will have an error term for which we compute a variance. That is, the variances of ϵ_7 and ϵ_8 represent the unexplained or residual variance in the two latent endogenous variables. We also have three structural paths: SES66 → Alien67, SES66 → Alien71, and Alien67 → Alien71. We estimate the variance of the exogenous latent variable SES66. Thus we have six unknowns for the structural part of our structural equation model. Altogether, we have $3+6+3+2+1 = 15$ parameters to estimate. With 21 possible equations, one for each variance and covariance, and 15 parameters to estimate, we have $21 - 15 = 6$ degrees of freedom. Thus we are able to fit this model even though there are only two indicators for each latent variable.

This counting rule is helpful, but it does not always work. It is a necessary but not a sufficient condition for identification. Our structural model is recursive, meaning that the causal flow only goes in one direction—there is no feedback. We can justify this assumption with the time ordering. SES66 is measured before either indicator of alienation, thus we can rule out that alienation in 1967 caused SES66. Similarly, we can also rule out that alienation in 1971 caused either SES66 or alienation in 1967. Can you see the possible problem if both alienation and SES were first measured in 1967? Alienation might not only depend on SES but also simultaneously cause SES. We might argue that people who were highly alienated would not be able to keep high occupational status jobs, and this could reduce their SES. We would have a reciprocal relationship between SES67 $\leftrightarrow$ Alien67. This would only use one more degree of freedom, but it would make our model nonrecursive.[1]

Box 3.1. Reordering labels for the error terms

Fitting the model does not depend on the order of the error term labels, but you may want to override the initial order Stata gives you. In figure 3.1, we first label the error terms for our indicators from ϵ_1 to ϵ_6, and then label the error terms for our latent endogenous variables ϵ_7 and ϵ_8.

To customize the label for an error term, click on the **Select** tool, and then double-click on the error term you want to reorder. In the dialog box that opens, you can specify a label in the *Custom label* box. Suppose the error term for Alien67 was ϵ_3, and we wanted to change it to ϵ_7. We can enter ϵ_7 as {&epsilon}{sub: 7}, and then click on **OK**. The ampersand, &, precedes the Greek letter. The sub: within braces makes whatever follows it a subscript. A sup: within braces would make whatever follows it a superscript.

By default, Stata uses the Greek letter ϵ ("epsilon") for all error terms. However, you might encounter an article that uses a δ ("delta") for error terms of indicators of exogenous latent variables, an ϵ for error terms of indicators of endogenous latent variables, and a ζ ("zeta") for error terms of the endogenous latent variables. You can imagine how we would do this. To label the error term for Alien67 as ζ_1, we would double-click on the error term for Alien67, and then we would enter {&zeta}{sub: 1} in the *Custom label* box.

---

1. We discussed nonrecursive path models in chapter 2. The same identification issues apply to nonrecursive full structural equation models.

3.2.2 Fitting a full structural equation model

For this example, we use a dataset from Stata's *Structural Equation Modeling Reference Manual*. These are not raw data but are data in the form of a covariance matrix. All we need to fit a structural equation model is a matrix of correlations and a vector of standard deviations.[2] We have three latent variables and need to specify measurements for each of them. We have three paths in our structural model that link the latent variables. Here is our program with the postestimation commands you learned in chapters 1 and 2.

```
. use http://www.stata-press.com/data/r13/sem_sm2.dta
. sem                                    ///
        (Alien67 -> anomia67 pwless67)   /// Measure Alien67
        (Alien71 -> anomia71 pwless71)   /// Measure Alien71
        (SES66 -> educ66 occstat66)      /// Measure SES66
        (Alien67 <- SES66)               /// Structural piece
        (Alien71 <- Alien67 SES66),      /// Structural piece
        standardized                     // Options
. estat eqgof                            // R-squared
. estat gof                              // Goodness of fit
. estat teffects                         // Direct, indirect, total
. estat mindices                         // Modification indices
```

When we are analyzing summary statistics data (a covariance matrix rather than raw data), we do not have the `method(mlmv)` option to handle missing data. The first three `sem` command lines above show the measurement models for each of the three latent variables, `Alien67`, `Alien71`, and `SES66`. You might notice that the direction of the arrow for all three is from the latent variable to the indicators. It does not matter whether the latent variable or the indicators come first as long as the arrow points toward the indicators.

The next two lines define the structural components of our model: `Alien67` ← SES66 and `Alien71` ← `Alien67` SES66. This reflects figure 3.1, where alienation in 1967 depends only on SES in 1966, while alienation in 1971 depends on both alienation in 1967 and SES in 1966. As in the measurement portion of the model, the ordering of the variables does not matter so long as the arrow points to the dependent variable. We finish our **sem** command with the standardized option because we want a standardized solution.

---

2. I discuss this data format in appendix B at the end of the text.

When you enter the `sem` command in Stata, you get the following results:

```
. sem (Alien67 -> anomia67 pwless67) (Alien71 -> anomia71 pwless71)
> (SES66 -> educ66 occstat66) (Alien67 <- SES66) (Alien71 <- Alien67 SES66),
> standardized
```

Endogenous variables

Measurement: anomia67 pwless67 anomia71 pwless71 educ66 occstat66
Latent: Alien67 Alien71

Exogenous variables

Latent: SES66

Fitting target model:

  (*output omitted*)

```
Structural equation model                    Number of obs      =        932
Estimation method  = ml
Log likelihood     = -15246.469

 ( 1)   [anomia67]Alien67 = 1
 ( 2)   [anomia71]Alien71 = 1
 ( 3)   [educ66]SES66 = 1
```

Standardized	Coef.	OIM Std. Err.	z	P>|z|	[95% Conf. Interval]	
Structural						
Alien67 <-						
SES66	-.5668218	.0344036	-16.48	0.000	-.6342517	-.4993919
Alien71 <-						
Alien67	.6630088	.0396724	16.71	0.000	.5852523	.7407654
SES66	-.151492	.0458162	-3.31	0.001	-.24129	-.061694
Measurement						
anomia67 <-						
Alien67	.812882	.0194328	41.83	0.000	.7747943	.8509697
_cons	3.95852	.097363	40.66	0.000	3.767692	4.149347
pwless67 <-						
Alien67	.811926	.0194466	41.75	0.000	.7738113	.8500406
_cons	4.796692	.1158294	41.41	0.000	4.56967	5.023713
anomia71 <-						
Alien71	.8395125	.0193263	43.44	0.000	.8016337	.8773913
_cons	3.993669	.09813	40.70	0.000	3.801338	4.186
pwless71 <-						
Alien71	.798082	.0198613	40.18	0.000	.7591546	.8370095
_cons	4.717723	.1140761	41.36	0.000	4.494137	4.941308
educ66 <-						
SES66	.8326718	.031738	26.24	0.000	.7704664	.8948772
_cons	3.518017	.0878219	40.06	0.000	3.345889	3.690145
occst~66 <-						
SES66	.6485148	.0301669	21.50	0.000	.5893887	.707641
_cons	1.767678	.0524337	33.71	0.000	1.66491	1.870446

var(e.anom~67)	.3392229	.0315932	.2826241	.4071562
var(e.pwle~67)	.3407762	.0315784	.2841788	.4086457
var(e.anom~71)	.2952187	.0324493	.2380034	.3661885
var(e.pwle~71)	.3630651	.0317019	.3059565	.4308333
var(e.educ66)	.3066577	.0528548	.2187474	.4298974
var(e.occs~66)	.5794285	.0391274	.5075984	.6614233
var(e.Alien67)	.6787131	.0390015	.6064191	.7596255
var(e.Alien71)	.4236057	.0345717	.360988	.4970851
var(SES66)	1	.	.	.

```
LR test of model vs. saturated: chi2(6)    =    71.62, Prob > chi2 = 0.0000
```

The standardized coefficients appear in figure 3.2. To create this figure, I used the SEM Builder and deleted some of the parameter estimates reported by default (the appendix to chapter 2 shows you how to delete these). The SEM Builder is described briefly in the appendices for chapters 1 and 2 and more fully in appendix A at the end of the book. If you are not familiar with the SEM Builder yet, you should work through all of appendix A quite carefully at this point.

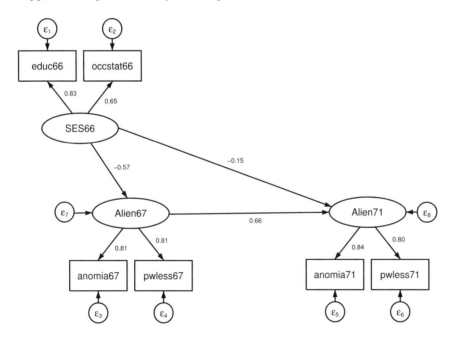

Figure 3.2. Stability of alienation ($N = 932$)

The initial structural equation model has 932 observations. We are told which three observed variables were used as reference indicators and therefore fixed at 1.0 for the unstandardized solution, even though we asked for a standardized solution.

```
( 1)   [anomia67]Alien67 = 1
( 2)   [anomia71]Alien71 = 1
( 3)   [educ66]SES66 = 1
```

When we ask for the standardized solution, both the latent and the observed variables are rescaled automatically to have a variance of 1.0, which allows us to have standardized estimates for each of the loadings for the measurement model. All six of these are strong, 0.65 to 0.84. These results are included in figure 3.2. The z tests show that all the loadings are statistically significant.

For the structural part of the model, we see that SES66 has a strong effect on Alien67: $\beta = -0.57$, $z = -16.48$, $p < 0.001$. Both SES66 and Alien67 have a significant effect on Alien71. The coefficient on the path from Alien67 to Alien71 is $\beta = 0.66$, $z = 16.71$, $p < 0.001$, and the coefficient on the path from SES66 to Alien71 is $\beta = -0.15$, $z = -3.31$, $p < 0.01$.

Box 3.2. Changing the reference indicator

Stata assumes the first observed variable used to measure each latent variable is the reference indicator. It normally does not matter which observed variable is the reference indicator; however, it makes sense to pick a strong indicator as the reference indicator. In the model shown in figure 3.2, the first indicators—educ66, anomia67, and anomia71—are the reference indicators, and each of them has the strongest standardized loading. When the first indicator is relatively weaker than the others, it makes sense to rearrange the variables so that the strongest indicator appears first. You might need to first fit the model with the default (the first indicator) as the initial reference indicator and then pick the indicator that has the largest loading as the replacement reference indicator.

To change the reference indicator, you simply list the indicator you want to be the reference first. For example, our sem command above has (SES66 -> educ66 occstat66). If we wanted to have occstat66 be the reference indicator, we would replace that portion of the command with (SES66 -> occstat66 educ66). Alternatively, we could still use our original sem command and simply constrain the coefficient on occstat66 to 1 by appending @1 to the variable name: (SES66 -> educ66 occstat66@1). Doing that would cause the unstandardized solution to have a fixed loading of 1.0 for occstat66.

Now we look at the results for the equation-level goodness-of-fit postestimation command to see how much variance we explain for each endogenous variable.

```
. estat eqgof
Equation-level goodness of fit
```

depvars	fitted	Variance predicted	residual	R-squared	mc	mc2
observed						
anomia67	11.8209	7.810982	4.009921	.6607771	.812882	.6607771
pwless67	9.353552	6.166084	3.187468	.6592238	.811926	.6592238
anomia71	12.51815	8.822558	3.695593	.7047813	.8395125	.7047813
pwless71	9.974882	6.35335	3.621531	.6369349	.798082	.6369349
educ66	9.599689	6.65587	2.943819	.6933423	.8326718	.6933423
occstat66	449.8053	189.1753	260.63	.4205715	.6485148	.4205715
latent						
Alien67	7.810982	2.509567	5.301416	.3212869	.5668218	.3212869
Alien71	8.822558	5.085272	3.737286	.5763943	.7592064	.5763943
overall				.7784845		

```
mc  = correlation between depvar and its prediction
mc2 = mc^2 is the Bentler-Raykov squared multiple correlation coefficient
```

We see that 32.1% of the variance in the latent variable `Alien67` and 57.6% of the variance in `Alien71` were explained. The pre-SEM literature suggested that alienation would not be stable; however, the stability coefficient for `Alien67` → `Alien71`, $\beta = 0.66$, is highly significant and substantial, especially given the time interval between its measurements in 1967 and 1971. The stability of alienation is substantially higher in our SEM approach because we specify anomia and powerlessness as measurements of alienation but also account for other variation in these variables through their corresponding error terms; that is, we include a measurement portion of the model in addition to the structural portion.

Next we look at the results of `estat gof, stats(all)` to see how well our model fits the data. The chi-squared statistic indicates that our model significantly fails to perfectly reproduce the original covariance matrix. Our model's $\chi^2(6) = 71.62$, $p < 0.001$, means that it fails to fully account for all the variances and covariances. The root mean squared error of approximation (RMSEA) of 0.11 is well above the goal of being less than 0.05. The comparative fit index (CFI) of 0.97, however, is better than our target of 0.95 for a good fit.

```
. estat gof, stats(all)
```

Fit statistic	Value	Description
Likelihood ratio		
chi2_ms(6)	71.621	model vs. saturated
p > chi2	0.000	
chi2_bs(15)	2134.080	baseline vs. saturated
p > chi2	0.000	
Population error		
RMSEA	0.108	Root mean squared error of approximation
90% CI, lower bound	0.087	
upper bound	0.131	
pclose	0.000	Probability RMSEA <= 0.05
Information criteria		
AIC	30534.938	Akaike's information criterion
BIC	30636.522	Bayesian information criterion
Baseline comparison		
CFI	0.969	Comparative fit index
TLI	0.923	Tucker-Lewis index
Size of residuals		
SRMR	0.019	Standardized root mean squared residual
CD	0.778	Coefficient of determination

A major goal of our structural equation model is to estimate the indirect effects, which we will report for our final model. We would not report these now because our model can still be improved to have a better fit to the data. To see how our model might be improved, we ask for modification indices. We have 6 degrees of freedom for our chi-squared, so there should be several ways to improve our fit.

```
. estat mindices
Modification indices
```

	MI	df	P>MI	EPC	Standard EPC
Measurement					
anomia67 <-					
anomia71	51.977	1	0.00	.3906425	.4019984
pwless71	32.517	1	0.00	-.2969297	-.2727609
educ66	5.627	1	0.02	.0935048	.0842631
pwless67 <-					
anomia71	41.618	1	0.00	-.3106995	-.3594367
pwless71	23.622	1	0.00	.2249714	.2323233
educ66	6.441	1	0.01	-.0889042	-.0900664
anomia71 <-					
anomia67	58.768	1	0.00	.429437	.4173061
pwless67	38.142	1	0.00	-.3873066	-.3347904

```
pwless71 <-
                anomia67        46.188       1    0.00    -.3308484    -.3601641
                pwless67        27.760       1    0.00     .2871709     .2780833

educ66 <-
                anomia67         4.415       1    0.04     .1055965     .1171781
                pwless67         6.816       1    0.01    -.1469371    -.1450411

cov(e.anomia67,e.anomia71)      63.786       1    0.00    1.951578      .5069627
cov(e.anomia67,e.pwless71)      49.892       1    0.00   -1.506704     -.3953794
     cov(e.anomia67,e.educ66)    6.063       1    0.01     .5527612     .1608845
cov(e.pwless67,e.anomia71)      49.876       1    0.00   -1.534199     -.4470094
cov(e.pwless67,e.pwless71)      37.357       1    0.00    1.159123      .341162
     cov(e.pwless67,e.educ66)    7.752       1    0.01    -.5557802    -.1814365
```

EPC = expected parameter change

The column labeled MI contains the modification indices, which are estimates of how much we can reduce the chi-squared for our model if we free individual parameters.[3] Several of the suggested modifications make no sense, so we need to be extremely cautious in using the modification indices to make changes. We only want to make a change in our model if it can be justified conceptually. For example, if we added a direct path from anomia71 going backward in time to anomia67, we would reduce chi-squared by approximately 51.98 points. With 1 degree of freedom, this would certainly be a significant improvement ($p < 0.001$), and the expected change in the standardized coefficient (which was fixed at 0.00 in our original model) would be about 0.40. Although this would improve our fit, it would make no sense in our model: a person's level of anomia in 1971 could not reasonably be expected to cause their level of anomia five years earlier.

3.2.3 Modifying our model

Let us focus on the modification indices for the covariances of the error terms. It would make sense to have the error terms for anomia67 and anomia71 be correlated as well as the error terms for pwless67 and pwless71. These changes in our model reflect the idea that unobserved variables are shared by these respective error terms. For example, psychological variables such as neuroticism are not in our model but are stable personality traits that might influence your sense of powerlessness at both waves of data collection. Although we do not observe these variables, we can allow for their influence by correlating the error terms, which is like acknowledging the existence of some level of spuriousness.[4]

---

3. Modification indices are based on freeing parameters one at a time, not sets of parameters.
4. Spuriousness means that a third variable explains the relationship between two variables. For example, neuroticism might explain part of the relationship between anomia67 and anomia71. Thus we could say that part of the relationship between anomia in 1967 and anomia in 1971 is spurious because of the common antecedent cause of neuroticism.

Figure 3.3 shows two models:[5] panel A has correlated error terms, and panel B shows unobserved latent variables.

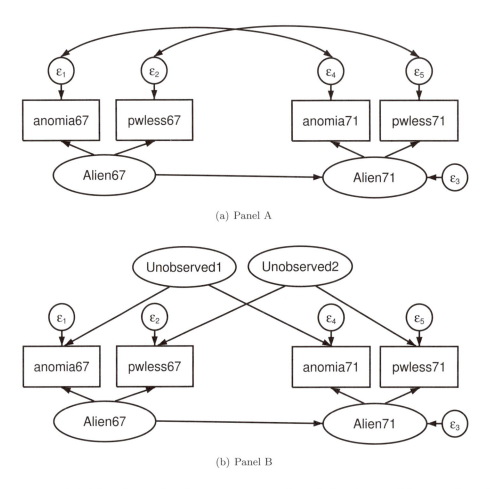

(a) Panel A

(b) Panel B

Figure 3.3. Model with correlated error terms to acknowledge the possible operation of unobserved variables

Adding a correlated error probably makes sense in this example, but we can anticipate important consequences. When the correlated errors are included, the coefficient on the path from Alien67 to Alien71 most likely will be reduced. This is because we are saying part of the relationship between the indicators of alienation is spurious because of a common (although unobserved) antecedent variable, as illustrated in figure 3.3 panel B. Without allowing for the errors to be correlated, we would have a

5. Both models exclude SES66 for simplification.

larger coefficient on the path from `Alien67` to `Alien71` but a relatively poor fit for our model. With the correlated errors, we will have weakened the coefficient on the path from `Alien67` to `Alien71` but will have a better fitting model.

In panel A, the covariance between `anomia67` and `anomia71` has two possible explanations. One is that both measures of anomia are caused by `Alien67` and `Alien71`, respectively, and these are causally related. The other is that the residuals for `anomia67` and `anomia71` are correlated. If this extra link is significant, then the path from `Alien67` to `Alien71` will not be as large. The same thing happens in panel B: if we add a variable (here called Unobserved1) that explained anomia at both waves, then the coefficient on the path from `Alien67` to `Alien71` might be reduced.

We are not able to fit the model in panel B because we do not have the necessary explanatory variables. However, the model in panel A makes considerable sense. By looking at the modification indices, we can see that estimating these correlated errors might help our fit—recognizing that once we estimate one of them, the modification indices for everything else will change. I think we can justify correlating the error terms for anomia at both waves and the error terms for powerlessness at both waves. These are not the biggest modification indices, but they are substantial and correlating them makes sense conceptually. Here we choose to free two correlated errors at once because it makes sense conceptually for both to be correlated; we have no conceptual basis to pick just one to be correlated.

How do our commands change? We need to add options to allow the error terms to be correlated. So far, we have not given these error terms names in our `sem` command. Stata automatically names them for us by adding an `e.` in front of the variable name. For example, the error term for `anomia67` is `e.anomia67`. We tell Stata to let the error terms covary. With a standardized solution, this adds an estimated correlation of the error terms to our model.

```
. sem                                        ///
      (Alien67 -> anomia67 pwless67)         /// Measure Alien67
      (Alien71 -> anomia71 pwless71)         /// Measure Alien71
      (SES66 -> educ66 occstat66)            /// Measure SES66
      (Alien67 <- SES66)                     /// Structural piece
      (Alien71 <- Alien67 SES66),            /// Structural piece
      cov(e.anomia67*e.anomia71)             /// Correlated error
      cov(e.pwless67*e.pwless71)             /// Correlated error
      standardized                          //  Standardized
. estat eqgof                               //  R-squared
. estat gof, stats(all)                     //  Goodness of fit
. estat teffects, nodirect standardized     //  Indirect effects
. estat mindices                            //  Modification indices
```

When we run this model, the goodness-of-fit measures (`estat gof, stats(all)`) are all excellent: $\chi^2(4) = 4.78$, $p = 0.31$; RMSEA = 0.01; and CFI = 1.00.

The standardized results, fit statistics, R^2's, and indirect effect of `SES66` on `Alien71` are shown in our final model (see figure 3.4). There are no modification indices reported. The default in `estat mindices` only reports modification indices greater than 3.84, corresponding to the 0.05 significance level with 1 degree of freedom.

3.2.4 Indirect effects

Indirect effects work in the same way that they do in a path model (see chapter 2). By examining figure 3.4, we see that `SES66` has an indirect effect on `Alien71` that is mediated by `Alien67` (`SES66` → `Alien67` → `Alien71`). We obtain the indirect effects with the postestimation command `estat teffects, nodirect standardized`. As explained in chapter 2, the `estat teffects` command provides the direct, indirect, and total effects. In our command, we added the option `nodirect` to specify that we do not need the direct effects (we have already estimated them).

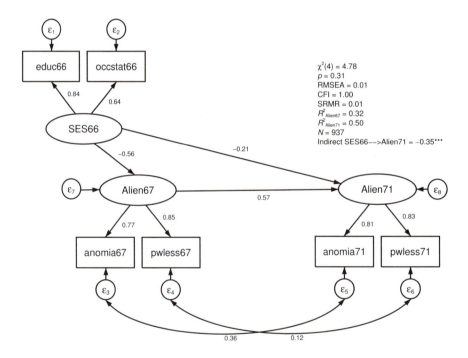

Figure 3.4. Final model (standardized; ***$p < 0.001$)

The standardized direct effect of `SES66` on `Alien71` is −0.21; thus the higher your `SES66`, the lower will be your level of alienation in 1971. This is as expected. However, `SES66` also directly influences your alienation in 1967, which in turn directly influences your alienation in 1971. Thus `SES66` indirectly influences `Alien71` as mediated by `Alien67`. Our command gives us an estimated indirect effect of −0.32; thus the indirect effect of `SES66` is larger than the direct effect. The z test of significance is for the unstandardized[6] indirect path, $z = -8.46$, $p < 0.001$.

We have only one indirect effect in this model, so these results are exactly what we need to know; in more complicated models, we may have two or more indirect effects for a single variable, a scenario we discussed in chapter 2. In box 2.3, you learned how to estimate and test specific indirect effects when they are part of your model. The `estat teffects` command does not give you an estimate of specific indirect effects automatically nor their significance when there are multiple indirect paths between a pair of variables.

You will notice that one of the correlated errors is substantial and significant, while the other is relatively small (0.12) and not significant. We leave them both in the model because we earlier decided on theoretical grounds that they should be included. Some researchers would argue that because we aim to develop a parsimonious model, we should rerun the model with only the significant correlated error term. The reality is that whether you leave a parameter with a small value in a model or remove it does not change much, simply because the parameter is already close to 0.

3.3   Equality constraints

When the same conceptual variable appears more than once as a latent variable in our model, we should consider using equality constraints. Before we can say that alienation in 1967 and alienation in 1971 are both measuring the same concept—that is, alienation—we need to establish the equivalence of the measurement of alienation (Raykov 1997a,b).

Let us imagine a more extreme example where we are correlating marital satisfaction of wives and their husbands. We have four indicators for each spouse: satisfaction with 1) sexual intimacy, 2) emotional support, 3) financial security, and 4) parenting performance of their spouse. Let us say that husbands put a huge emphasis on sexual intimacy (loading 0.80) and the parenting performance of their wives (loading 0.70), while putting less emphasis on financial security (loading 0.40) and emotional support (loading 0.50). Wives put greater emphasis on emotional support (loading 0.90) and financial security (loading 0.70), while putting less emphasis on sexual intimacy (loading 0.40) and the parenting performance of their husbands (loading 0.50).

---

6. A test of significance for the standardized indirect effect is not available directly from `estat teffects`. To obtain tests of standardized indirect effects, you can use the method described in chapter 2 (see box 2.1).

If this were the case, then marital satisfaction would have a fundamentally different meaning for a husband than it has for his wife. His view of marriage would put the emphasis on sexual satisfaction and his wife's being a good mother, while her view of marriage would emphasize emotional support and financial security. Such a model with hypothetical coefficients appears in figure 3.5.

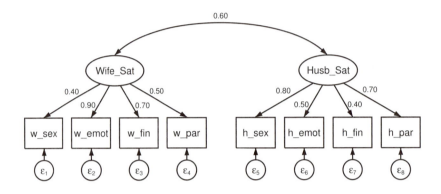

Figure 3.5. A model of marital satisfaction (hypothetical data)

Before SEM, researchers routinely generated a total or average score for the indicators and correlated these. This method assumes that each source of marital satisfaction counts equally for both a husband and his wife. At the least, a structural equation model analysis allows for differential importance on different aspects of marital satisfaction.

Equality constraints let us extend this improvement over traditional summated scales not only to allow for differential importance of different aspects, but also to test whether this applies equally for wives and their husbands. Does sexual intimacy have equal importance for wives and husbands? Does emotional support have equal importance for wives and husbands? If this is not the case (and it is not; see figure 3.5), then marital satisfaction has a different meaning for a wife than it does for her husband, and we need to be aware of this.

The same idea applies to the example of alienation in 1967 and alienation in 1971. The meaning of alienation may change to have greater importance for either powerlessness or anomia at one wave than at the other wave. It may be that in 1967, powerlessness was less important than anomia, but vice versa in 1971. If that happened, then we would need to acknowledge that these concepts may go by the same name, but they have different meanings for our study participants. The meaning of alienation would vary from one wave to the next.

There are four widely recognized levels of invariance, which are detailed in table 3.1. The first and lowest level requires that the same set of indicators be relevant to the latent variable. The hypothetical results in figure 3.5 meet this level of invariance. In our example about alienation (figure 3.4), powerlessness might have been more central in 1967 than it was in 1971. This first level allows both the loadings and the error variances for the observed measures to vary from one wave to the next. There may be more or less unique variance for `e.anomia67` or `e.pwless67` than for `e.anomia71` or `e.pwless71`, respectively. If this is the best case you can make for invariance, you need to make this limitation clear to your reader; to some extent, we are correlating apples with oranges.

The second level of invariance might be called partial invariance. Like in the first level, all the indicators are included; the partial invariance adds the requirement that some of the indicators have the same loadings and others differ only by a fairly small amount. As you can imagine, there is considerable subjectivity in deciding what level of similarity of corresponding loadings is sufficient. If you use this level of invariance, you need to acknowledge this limitation to your reader.

The third level of invariance is the next step up for measuring equivalence, and this is the most widely used level for structural equation models. This level requires the corresponding loadings to be invariant but allows the error variances to be free. This way, each latent concept is measured with consistency (same corresponding loadings) but with different error variances allowed.

The fourth level is the most restrictive invariance measurement. This level is achieved when both the loadings and the error variances are invariant across waves. This is an ideal level for SEM because it ensures that you are correlating like with like. Because this level of invariance is rarely achieved, most SEM analysts are satisfied if they have level three invariance.

Table 3.1. Requirements for levels of invariance

Level of invariance	All loadings significant	Similar loadings	Loadings equal	Equal error variance
1—Same form	Yes	No	No	No
2—Similar loadings	Yes	Yes	No	No
3—Invariant loadings	Yes	Yes	Yes	No
4—Inv. loadings and errors	Yes	Yes	Yes	Yes

3.4 Programming constraints

We should run tests of invariance on the unstandardized model. We have been fitting standardized models because that is most common for structural equation models.

However, the standardized coefficients depend on both the unstandardized values and the relative variances of the variables. Perhaps the unstandardized values are the same. In this case, the corresponding standardized values will be the same only if the variables have identical variances. Because the standardized solution mixes the unstandardized effects and the relative variances, it is usual to compare coefficients for the unstandardized results. The unstandardized solution measures the structural relationships; the standardized solution transforms the unstandardized solution by forcing all variables, both observed and latent, to have identical variances of 1.0.

The first step to assess equality is to obtain an unstandardized solution. We accomplish this simply by dropping the `standardized` option from our command. If all the loadings are statistically significant, then we demonstrate level one invariance—same form. The following results for the unstandardized measurement model demonstrate that we have at least level one invariance:

(output omitted)

Measurement						
anomia67 <-						
Alien67	1	(constrained)				
_cons	13.61	.1126143	120.85	0.000	13.38928	13.83072
pwless67 <-						
Alien67	.9785952	.0619825	15.79	0.000	.8571117	1.100079
_cons	14.67	.1001814	146.43	0.000	14.47365	14.86635
anomia71 <-						
Alien71	1	(constrained)				
_cons	14.13	.1159036	121.91	0.000	13.90283	14.35717
pwless71 <-						
Alien71	.9217508	.0597225	15.43	0.000	.8046968	1.038805
_cons	14.9	.1034517	144.03	0.000	14.69724	15.10276
educ66 <-						
SES66	1	(constrained)				
_cons	10.9	.1014894	107.40	0.000	10.70108	11.09892
occst~66 <-						
SES66	5.22132	.425595	12.27	0.000	4.387169	6.055471
_cons	37.49	.6947112	53.96	0.000	36.12839	38.85161

(output omitted)

We see that the observed variables we picked as reference indicators now have fixed loadings of 1.0. Though we do not need a test of significance for `anomia67` and `anomia71` because they have identical fixed values, it is useful to check that the standardized loadings for both of them are significant. Both anomia and powerlessness loadings were significant for the standardized solution. The loadings for anomia were automatically constrained to be equal to 1.0 for the unstandardized solution, while the loadings for powerlessness are 0.98, $z = 15.79$, $p < 0.001$ in 1967 and 0.92, $z = 15.43$, $p < 0.001$ in 1971. We have met the minimal requirements of level one invariance.

Can we meet the additional requirements of level three invariance, namely, that the corresponding loadings are not significantly different? With only two indicators at each wave and with one of them being the reference indicator, we only have to test the equivalence of the loadings for `pwless67` and `pwless71`. If we had three indicators, we would have to test the equivalence of two pairs of them; if we had four indicators, we would have to test the equivalence of three pairs of them (see box 3.3 below). If we were testing the equivalence of more than one pair of indicators and found that some but not all pairs met the criteria for level three invariance, we might still have level two invariance if other loadings were close in value. Because we are only testing one pair of indicators, we will not evaluate the possibility of level two invariance.

There are two strategies we can use. For the first approach, we can test whether the loadings for `pwless67` and `pwless71` are significantly different by using the `test` command that we learned about in chapter 2. To run this command, we need to know the nicknames that Stata has assigned to the different parameters. After running our unstandardized `sem` command, we can obtain the nicknames by running `sem, coeflegend`.

```
. sem, coeflegend
Structural equation model                    Number of obs      =      932
Estimation method   = ml
Log likelihood      = -15213.046
 ( 1)   [anomia67]Alien67 = 1
 ( 2)   [anomia71]Alien71 = 1
 ( 3)   [educ66]SES66 = 1
```

	Coef.	Legend
Structural		
Alien67 <-		
SES66	-.5752228	_b[Alien67:SES66]
Alien71 <-		
Alien67	.606954	_b[Alien71:Alien67]
SES66	-.2270301	_b[Alien71:SES66]
Measurement		
anomia67 <-		
Alien67	1	_b[anomia67:Alien67]
_cons	13.61	_b[anomia67:_cons]
pwless67 <-		
Alien67	.9785952	_b[pwless67:Alien67]
_cons	14.67	_b[pwless67:_cons]
anomia71 <-		
Alien71	1	_b[anomia71:Alien71]
_cons	14.13	_b[anomia71:_cons]
pwless71 <-		
Alien71	.9217508	_b[pwless71:Alien71]
_cons	14.9	_b[pwless71:_cons]

```
       educ66 <-
           SES66           1   _b[educ66:SES66]
           _cons         10.9  _b[educ66:_cons]

      occst~66 <-
           SES66      5.22132  _b[occstat66:SES66]
           _cons        37.49  _b[occstat66:_cons]

  var(e.anom~67)    4.728874  _b[var(e.anomia67):_cons]
  var(e.pwle~67)    2.563413  _b[var(e.pwless67):_cons]
  var(e.anom~71)    4.396081  _b[var(e.anomia71):_cons]
  var(e.pwle~71)    3.072085  _b[var(e.pwless71):_cons]
   var(e.educ66)    2.803674  _b[var(e.educ66):_cons]
  var(e.occs~66)   264.5311   _b[var(e.occstat66):_cons]
  var(e.Alien67)    4.842059  _b[var(e.Alien67):_cons]
  var(e.Alien71)    4.084249  _b[var(e.Alien71):_cons]
      var(SES66)    6.796014  _b[var(SES66):_cons]

  cov(e.anom~67,
     e.anomia71)    1.622024  _b[cov(e.anomia67,e.anomia71):_cons]
  cov(e.pwle~67,
     e.pwless71)     .3399961 _b[cov(e.pwless67,e.pwless71):_cons]
```

LR test of model vs. saturated: chi2(4)   =   4.78, Prob > chi2 = 0.3111

We see that _b[*name*] is the internal name assigned to each of the parameter estimates. To test whether the loading for Alien67 → pwless67 equals the loading for Alien71 → pwless71, we need to test for a difference between _b[pwless67: Alien67] and _b[pwless71: Alien71].

```
. test (_b[pwless67:Alien67]==_b[pwless71:Alien71])
 ( 1)  [pwless67]Alien67 - [pwless71]Alien71 = 0
          chi2(  1) =     0.89
        Prob > chi2 =     0.3464
```

This is a Wald $\chi^2(1) = 0.89$, $p = 0.35$, which is not significant. Thus there is not a significant difference between these parameter estimates. Because the loadings for anomia67 and anomia71 were both fixed at 1.0 and the difference between the loadings for pwless67 and pwless71 is not significant, we have met the requirements of level three invariance.

Box 3.3. Testing simultaneous equality constraints with three or more indicators

Suppose we have three indicators for a pair of latent variables, marital satisfaction for wives and marital satisfaction for their husbands. Our indicators for the husband are x1h = sexual intimacy, x2h = emotional support, and x3h = division of household chores; the same set of indicators for the wife are x1w, x2w, and x3w. We might make x1h and x1w the reference indicators, which means that we need a simultaneous test for x2h = x2w and x3h = x3w. We run sem, coeflegend to learn the internal names Stata creates: _b[x2h: Husb_sat], _b[x3h: Husb_sat], _b[x2w: Wife_sat], and _b[x3w: Wife_sat]. Our test command would need to simultaneously test two equalities:

```
. test (_b[x2h: Husb_sat] = _b[x2w: Wife_sat]) ///
(_b[x3h: Husb_sat] = _b[x3w: Wife_sat])
```

The second approach to testing for level three invariance is to fit two structural equation models. We have already fit the first model where $\chi^2(4) = 4.78$, $p = 0.31$. The second model is identical except that we place constraints so the loadings for Alien67 → pwless67 and Alien71 → pwless71 must have identical values (note we are working with the unstandardized result).

One way to fit our model is to attach the same label to each of these paths; we will use @a1 as our label. Every parameter with this label attached to it will be forced by Stata to have identical estimated values. Thus because pwless67@a1 and pwless71@a1 have the same label, Stata solves for the best maximum likelihood solution with the constraint that these two unstandardized loadings are identical:

```
. sem (Alien67 -> anomia67 pwless67@a1)        /// Measure Alien67
>     (Alien71 -> anomia71 pwless71@a1)        /// Measure Alien71
>     (SES66 -> educ66 occstat66)              /// Measure SES66
>     (Alien67 <- SES66)                       /// Structural piece
>     (Alien71 <- Alien67 SES66),              /// Structural piece
>     cov(e.anomia67*e.anomia71)               /// Correlated error
>     cov(e.pwless67*e.pwless71)               //  Correlated error
  (output omitted )
Structural equation model                    Number of obs     =       932
Estimation method   = ml
Log likelihood      = -15213.491
 ( 1)   [anomia67]Alien67 = 1
 ( 2)   [pwless67]Alien67 - [pwless71]Alien71 = 0
 ( 3)   [anomia71]Alien71 = 1
 ( 4)   [educ66]SES66 = 1
```

	Coef.	OIM Std. Err.	z	P>|z|	[95% Conf. Interval]	
Structural						
Alien67 <-						
SES66	-.5846697	.0573663	-10.19	0.000	-.6971055	-.4722339
Alien71 <-						
Alien67	.5911718	.0475116	12.44	0.000	.4980508	.6842928
SES66	-.2221104	.0525345	-4.23	0.000	-.3250762	-.1191447
Measurement						
anomia67 <-						
Alien67	1	(constrained)				
_cons	13.61	.1129989	120.44	0.000	13.38853	13.83147
pwless67 <-						
Alien67	.951234	.052855	18.00	0.000	.8476401	1.054828
_cons	14.67	.0997932	147.00	0.000	14.47441	14.86559
anomia71 <-						
Alien71	1	(constrained)				
_cons	14.13	.1155494	122.29	0.000	13.90353	14.35647
pwless71 <-						
Alien71	.951234	.052855	18.00	0.000	.8476401	1.054828
_cons	14.9	.1038614	143.46	0.000	14.69644	15.10356
educ66 <-						
SES66	1	(constrained)				
_cons	10.9	.1014894	107.40	0.000	10.70108	11.09892
occst~66 <-						
SES66	5.228402	.4258885	12.28	0.000	4.393676	6.063129
_cons	37.49	.6947112	53.96	0.000	36.12839	38.85161
var(e.anom~67)	4.61177	.442892			3.82052	5.566894
var(e.pwle~67)	2.686334	.3774104			2.039729	3.537918
var(e.anom~71)	4.558759	.4808796			3.707301	5.605772
var(e.pwle~71)	2.918973	.4127779			2.212381	3.851237
var(e.educ66)	2.81288	.5107798			1.970538	4.015296
var(e.occs~66)	264.2798	18.22499			230.8682	302.5266
var(e.Alien67)	4.968713	.4535475			4.154757	5.94213
var(e.Alien71)	3.96083	.3684007			3.300769	4.752885
var(SES66)	6.786808	.6516849			5.622523	8.192188
cov(e.anom~67, e.anomia71)	1.627928	.3153949	5.16	0.000	1.009765	2.24609
cov(e.pwle~67, e.pwless71)	.3284112	.2626751	1.25	0.211	-.1864225	.8432449

LR test of model vs. saturated: chi2(5)   =   5.66, Prob > chi2 = 0.3403

Near the top of the results, we see a list of the four constraints on our structural equation model:

```
( 1)  [anomia67]Alien67 = 1
( 2)  [pwless67]Alien67 - [pwless71]Alien71 = 0
( 3)  [anomia71]Alien71 = 1
( 4)  [educ66]SES66 = 1
```

The second listed constraint requests that there be no difference between the loading for `pwless67` and the loading for `pwless71`. Both of these loadings are constrained to be equal, and the estimated unstandardized value for both of them is 0.95.

(output omitted)

pwless67 <-						
Alien67	.951234	.052855	18.00	0.000	.8476401	1.054828
_cons	14.67	.0997932	147.00	0.000	14.47441	14.86559

(output omitted)

pwless71 <-						
Alien71	.951234	.052855	18.00	0.000	.8476401	1.054828
_cons	14.9	.1038614	143.46	0.000	14.69644	15.10356

(output omitted)

We can also run `estat gof, stats(chi2 rmsea indices)` so that we can compare the goodness-of-fit statistics for the first model (no constrained loadings) with those for the second model (loadings constrained to be equal). Here are the results for the second model:

```
. estat gof, stats(chi2 rmsea indices)
```

Fit statistic	Value	Description
Likelihood ratio		
chi2_ms(5)	5.664	model vs. saturated
p > chi2	0.340	
chi2_bs(15)	2134.080	baseline vs. saturated
p > chi2	0.000	
Population error		
RMSEA	0.012	Root mean squared error of approximation
90% CI, lower bound	0.000	
upper bound	0.048	
pclose	0.960	Probability RMSEA <= 0.05
Baseline comparison		
CFI	1.000	Comparative fit index
TLI	0.999	Tucker-Lewis index

In table 3.2, we compare the model with level one invariance where the indicators are the same with the model with level three invariance where the unstandardized loadings are the same. We test for a significant difference and find none: $\chi^2(1) = 0.88$, $p = 0.35$. To obtain the probability value in Stata, you can use the following density function:

```
. display "p = " chi2tail(1,0.88)
p = .34820168
```

Table 3.2. Comparison of two models

Model	Chi-squared	df	p	RMSEA	CFI
Level 1 (same indicators)	4.78	4	0.31	0.01	1.00
Level 3 (invariant loadings)	5.66	5	0.34	0.01	1.00
Difference	0.88	1	0.35		

Because the chi-squared for the difference is not statistically significant, we conclude that the level three invariance is not worse than the level one invariance, and we select the level three model over the level one model. The comparison of nested models like this provides more information than the simple testing of the equality. In addition to a likelihood ratio chi-squared test based on the difference of the two chi-squared values and the difference in the number of degrees of freedom, we also have a comparison of the measures of goodness of fit. As table 3.2 shows, both models have CFIs of 1.0, and the difference in RMSEAs is minimal (before rounding). It is entirely appropriate to fit the second model and assume that the meaning of alienation is invariant over time.

When the unstandardized solution has equal loadings, the standardized solution will not have identical values for loadings unless both observed variables have precisely the same variance. If we run the last model with exactly the same command except adding the **standardized** option, we would get a loading of 0.843 for 1967 and 0.842 for 1971. This difference is not important because we establish level three invariance with the unstandardized solution.

You have learned how to fit two models and compare the chi-squared values. Stata has a nice way to automate this process by using the **estimates store** and **lrtest** commands. These work for a maximum likelihood solution because the chi-squared for the differences (0.88 in table 3.2) is a simple function of the likelihood ratio for the two models. After running the first model, we add the command **estimates store level1** (we could assign any name here; I chose **level1** because that was our least restrictive model). This command saves all the estimated values for this model. Then after running the second model, we add the command **estimates store level3** (named **level3** because we require the loadings to be invariant). Finally, we run **lrtest level1 level3**.

```
. sem (Alien67 -> anomia67 pwless67)        /// Measure Alien67
>     (Alien71 -> anomia71 pwless71)        /// Measure Alien71
>     (SES66 -> educ66 occstat66)            /// Measure SES66
>     (Alien67 <- SES66)                    /// Structural piece
>     (Alien71 <- Alien67 SES66),           /// Structural piece
>     cov(e.anomia67*e.anomia71)            /// Correlated error
>     cov(e.pwless67*e.pwless71)            //  Correlated error
  (output omitted)
. estimates store level1
. estat gof, stats(chi2 rmsea indices)
  (output omitted)
. sem (Alien67 -> anomia67 pwless67@a1)     /// Measure Alien67
>     (Alien71 -> anomia71 pwless71@a1)     /// Measure Alien71
>     (SES66 -> educ66 occstat66)            /// Measure SES66
>     (Alien67 <- SES66)                    /// Structural piece
>     (Alien71 <- Alien67 SES66),           /// Structural piece
>     cov(e.anomia67*e.anomia71)            /// Correlated error
>     cov(e.pwless67*e.pwless71)            //  Correlated error
  (output omitted)
. estimates store level3
. estat gof, stats(chi2 rmsea indices)
  (output omitted)
. lrtest level1 level3
Likelihood-ratio test                        LR chi2(1)  =       0.89
(Assumption: level3 nested in level1)        Prob > chi2 =     0.3457
```

We produced essentially the same values for the chi-squared difference; the slight difference is because lrtest keeps more decimal places in the calculations.

Box 3.4. Equality constraints using the SEM Builder

Once you have drawn the model, it is easy to add equality constraints with the SEM Builder. In the text above, you learned how to assign the same label to parameters you are constraining. We want the coefficient on the path from `Alien67` → `pwless67` to equal the coefficient on the path from `Alien71` → `pwless71`. Click on the **Select** tool, and then click on one of these paths. In the **Constrain parameters** box, 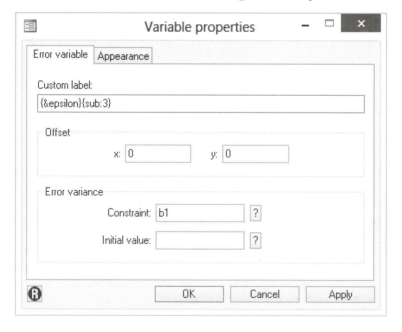 , on the toolbar, enter `a1` to attach the label `a1` to this path. Then click on the other path and enter the same constraint. We can now fit the unstandardized result and get the same results we got above.

We can also put equality constraints on error variances. For a model with level four invariance, we need to show that the corresponding variances for the error terms are equal; that is, we need to show that the variance of ϵ_3 for `anomia67` equals the variance of ϵ_5 for `anomia71`, and that the variance of ϵ_4 for `pwless67` equals the variance of ϵ_6 for `pwless71`. We do this in the same way that we put an equality constraint on the two loadings: click on ϵ_3 and enter `b1` in the **Constrain parameters** box, σ^2 , on the toolbar. Alternatively, you can double-click on the error itself and enter the constraint in the dialog box that opens:

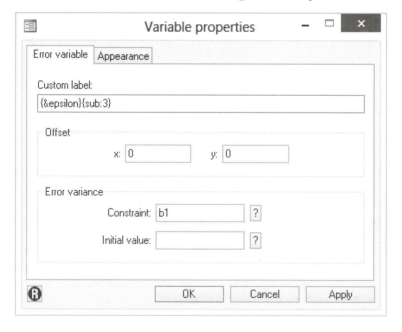

We also label ϵ_5 with `b1`. To separately constrain ϵ_4 and ϵ_6, we label them with `b2`.

Many researchers find the level four invariance model, requiring both the loadings and the error variances to be equal, to be too restrictive; most are satisfied with the level three model. After all, the level three model assumes that the indicators have the same salience (loadings) at both waves. This allows for differences in the unique variances. For the level four model in figure 3.4, we need to constrain the variance of ϵ_3 to be equal to the variance of ϵ_5, and constrain the variance of ϵ_4 to be equal to the variance of ϵ_6. Box 3.4 shows how to do this using the SEM Builder. The `sem` command to accomplish this is as follows:

```
. sem (Alien67 -> anomia67 pwless67@a1)                        ///
      (Alien71 -> anomia71 pwless71@a1)                        ///
      (SES66 -> educ66 occstat66) (SES66 -> Alien67)           ///
      (SES66 Alien67 -> Alien71),                              ///
      cov(e.anomia67@b1 e.anomia71@b1 e.pwless67@b2 e.pwless71@b2 /// Constrained
      e.pwless67*e.pwless71 e.anomia67*e.anomia71) standardized  // Errors
```

In this command, we have indicated in the `cov()` option which error variances are constrained to be equal and which are allowed to covary. Both indicators of anomia are constrained to have the same error variance, as noted by the `b1` label for both. Both indicators of powerlessness are similarly constrained to have the same error variance, as noted by the `b2` label. Finally, the corresponding error terms at the two waves are free to covary.

You can fit the model with level four invariance, where both loadings and error variances are constrained to be equal, as an exercise. The level four model compares with the level three model in the same way we compared the level three model with the level one model (see table 3.2). If the chi-squared difference test does not show that the chi-squared values are significantly different, then we go with the more restricted model, that is, the level four invariance model. If the p-value for the chi-squared difference test is small enough to indicate a statistically significant difference in the two chi-squared values, then we compare relative fit by using the RMSEA and CFI tests.

3.5   Structural model with formative indicators

Most applications of SEM involve what are sometimes referred to as reflective indicators, where the latent variable causes the response on the observed variable. This way of measuring the latent variable is known as a reflective measurement model. People who are extremely alienated will express this in how they respond to questions about their powerlessness. The individuals' scores on anomia and powerlessness will, in turn, reflect their alienation. People who are not alienated will respond very differently to the same questions than will people who are alienated. The key is that the causal flow goes from the latent variable to the indicators. In this sense, the indicators reflect the latent variable, and so we say the observed indicator variables are reflective. This causal direction is sensible for most social psychological variables.

Because each indicator variable reflects the underlying latent variable, we expect the indicators to be at least moderately correlated with each other. Ideally, the indicators of a latent variable will be more correlated with one another than with indicators of other variables. A simple exploratory factor analysis should have all the reflective indicators load strongly on a single first dimension.

Other latent variables have causal flow in the opposite direction, that is, the latent variables are caused by the observed indicators. This type of latent variable is called a formative construct or a composite latent variable. A model with a composite latent variable is known as a formative measurement model.

Sometimes, a group of variables forms an index. For example, an index of juvenile delinquency might include a checklist of specific experiences: Have you stolen something in the last month? Have you been drunk in the last month? Have you skipped school in the last month? Have you cut someone in the last month? Have you used pot in the last month? Have you brought a gun to school? A person who checks several of these items with a yes response is considered highly delinquent. We think of their delinquency as a composite of how they respond to the set of items. Each item provides a largely independent piece of information to help us predict who is highly delinquent and who is not.

These items will vary in how correlated they are. For example, the correlation between using pot and cutting somebody is very low. In fact, we will improve our index if we pick a diverse set of dimensions of delinquency, which means that many of the correlations will be very low. A simple exploratory factor analysis will yield several factors.

Can you think of other examples of formative indicators for latent variables? What about the latent variable stress? We could conceptualize stress having either formative or reflective indicators. If we have a series of observed items that have highly correlated responses and that form a scale, we have reflective indicators. We might say that people who are very stressed will score higher on all the scale items, and the items will be highly correlated with each other. By contrast, if we have a checklist of possible stressful events (death of spouse, loss of job, bad grade in school, etc.), we have an index. We do not expect these items to be highly correlated; you can be stressed for many different and independent reasons. In this case, we have formative indicators.

Socioeconomic indicators can also be thought of as formative. In our model of alienation in 1967 and 1971, we have been assuming that SES66 causes your education and occupational status (reflective indicators). It surely makes more sense to reverse this and say that your education and occupational status cause your SES66. Imagine a 40-year-old man who gets a big promotion in a large company. His new occupational status will raise his SES66, therefore occstat66 → SES66. You would not say that his SES66 increased his occupational status. Similarly, if you had a 50-year-old woman who had high SES66, you would not say that her SES66 caused her completion of a PhD 22 years earlier. It makes much more sense to think of SES66 as a composite latent variable. In this example, the indicators are likely to be highly correlated with one another, but this is not essential for formative indicators.

3.5.1   Identification and estimation of a composite latent variable

Our model of the stability of education could be revised as shown in figure 3.6 to have formative indicators of SES66. We have done two things to identify this model:

- For the composite latent variable, we have fixed one of our formative indicators, educ66, to have a loading of 1.0. We should pick one of our most central indicators as the reference indicator.

- We have fixed the error variance for the composite latent variable, ϵ_7, to be 0.0.

We will carry over the equality constraint to Alien67 → pwless67 and Alien71 → pwless71, although this has nothing to do with our composite latent variable, SES66.

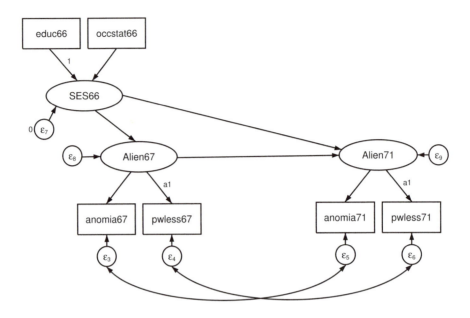

Figure 3.6. Formative model using the SEM Builder

The **sem** command is as follows:

```
.. sem                                        ///
>           (Alien67 -> anomia67 pwless67@a1) /// Measure Alien67
>           (Alien71 -> anomia71 pwless71@a1) /// Measure Alien71
>           (SES66 <- educ66@1 occstat66)     /// Formative indicators
>           (Alien67 <- SES66)                /// Structural piece
>           (Alien71 <- Alien67 SES66),       /// Structural piece
>           var(e.SES66@0)                    /// Fix error variance for SES66
>           cov(e.anomia67*e.anomia71)        /// Correlated error
>           cov(e.pwless67*e.pwless71)        /// Correlated error
>           standardized
   (output omitted)
Structural equation model                 Number of obs      =       932
Estimation method  = ml
Log likelihood     = -15213.491
  ( 1)  [anomia67]Alien67 = 1
  ( 2)  [pwless67]Alien67 - [pwless71]Alien71 = 0
  ( 3)  [anomia71]Alien71 = 1
  ( 4)  [SES66]educ66 = 1
  ( 5)  [var(e.SES66)]_cons = 0
```

Standardized	Coef.	OIM Std. Err.	z	P>|z|	[95% Conf. Interval]	
Structural Alien67 <- SES66	-.4908607	.0277376	-17.70	0.000	-.5452254	-.4364959

Alien71 <-
  Alien67
  SES66
	.6056019
	-.161013
		.0342491
		.0420615
			17.68
			-3.83
				0.000
				0.000
					.5384749
					-.2434519
						.6727289
						-.078574

SES66 <-
  educ66
  occstat66
	.8015395
	.3053268
		.0533007
		.0697827
			15.04
			4.38
				0.000
				0.000
					.697072
					.1685552
						.906007
						.4420983

Measurement
anomia67 <-
  Alien67
  _cons
	.7826057
	5.235835
		.0235488
		.1157068
			33.23
			45.25
				0.000
				0.000
					.7364509
					5.009054
						.8287605
						5.462616

pwless67 <-
  Alien67
  _cons
	.8429537
	6.205366
		.0241527
		.1277111
			34.90
			48.59
				0.000
				0.000
					.7956152
					5.955057
						.8902922
						6.455676

anomia71 <-
  Alien71
  _cons
	.7960215
	5.231156
		.0241664
		.1186566
			32.94
			44.09
				0.000
				0.000
					.7486563
					4.998593
						.8433868
						5.463718

pwless71 <-
  Alien71
  _cons
	.8424138
	5.99619
		.0241605
		.1273971
			34.87
			47.07
				0.000
				0.000
					.7950601
					5.746497
						.8897674
						6.245884

var(e.anom~67)	.3875284	.0368588			.3216203	.4669426
var(e.pwle~67)	.289429	.0407193			.2196787	.3813258
var(e.anom~71)	.3663497	.0384739			.2981968	.4500791
var(e.pwle~71)	.290339	.0407062			.2205794	.3821606
var(e.Alien67)	.7590558	.0272306			.7075179	.8143479
var(e.Alien71)	.5115937	.0329991			.450838	.5805371
var(e.SES66)	0	(constrained)				
cov(e.anom~67, e.anomia71)	.3550406	.047101	7.54	0.000	.2627244	.4473568
cov(e.pwle~67, e.pwless71)	.1172796	.0824791	1.42	0.155	-.0443765	.2789357

LR test of model vs. saturated: chi2(5) =   5.66, Prob > chi2 = 0.3403

The measurement piece for SES66 has the arrow reversed. Rather than SES66 → educ66 occstat66, we now have SES66 ← educ66@1 occstat66. Because these are formative rather than reflective indicators, we need to tell Stata to fix one of the formative indicators to have a loading of 1.0, which we do with the @1 part of the command. To fix the error variance for SES66 at 0.0, we add a line to the command, var(e.SES66@0). This is all we have to do. We can add the usual postestimation commands as appropriate.

Most of the results are familiar. Notice that SES66 is no longer part of the measurement model. SES66 moved to the structural model results section because it is now an endogenous variable explained by educ66 and occstat66. Because we asked for a standardized solution, we have a separate standardized loading and test of significance for both educ66 and occstat66. Stata accomplishes this by rescaling the unstandardized solution, which makes the variance of all variables, including SES66, equal to 1.0; thus we do not need a reference indicator to estimate its variance. As before, we have allowed the respective error terms for anomia and powerlessness to be correlated. And the last item of note: the variance of the error term for SES66, var(e.SES66), is now 0.0.

To fit this model using the SEM Builder, we need to add the indicated constraints. After placing the variables and appropriate paths in the path diagram, we use the **Select** tool to click on the path educ66 → SES66, and then we enter 1 in the **Constrain parameter** box in the toolbar. Next we constrain the loadings of pwless67 and pwless71 to be equal by clicking on each in turn and entering the constraint a1 for both of them. Finally, we click on the error for SES66, ϵ_7, and add the constraint of 0 for its variance. The standardized results are shown in figure 3.7.

When we run the postestimation command estat teffects, standardized, we now get indirect effects of education and occupational status on alienation in 1967 and alienation in 1971. For example, education influences alienation in 1971 as follows: educ66 → SES66 → Alien67 → Alien71 and educ66 → SES66 → Alien71.

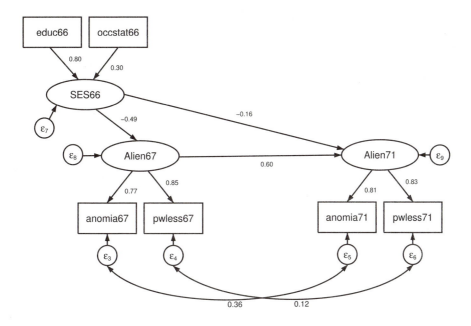

Figure 3.7. Standardized solution when SES66 is a composite variable

3.5.2   Multiple indicators, multiple causes model

What does it mean to fix the error variance of SES66 at 0.0 and the unstandardized coefficient from educ66 to SES66 at 1.0? This means our latent variable is assumed to be a perfect composite of the observed formative indicators. These two changes allow us to identify a formative construct.

Constraining the error variance to be 0.0 means that there is nothing more to SES66 than the effect of education and occupational status. This may be a difficult assumption to justify to your reader. What about income? A person who has limited education and limited occupational status may still have a reasonably high SES66 because he or she has a lot of money. An old HBO program called the Sopranos was about a leader of an organized gang who made a lot of money. If you did not happen to know how he made his income, you would think of him as having high SES66 independent of his education and the low status his occupation had.

To justify that our composite measure of SES66 has no error variance, we would want to include more than just education and occupational status. At the least, we would want to add income. Still, if we are willing to have the meaning of the composite latent variable be limited to the formative indicators we have available, then fixing the error variance at 0.0 is okay. This simplifies the identification of our model.

If we want a formative measurement model that includes a nonzero error variance for the latent variable, our model identification is more complex. We might achieve this by using what is called a multiple indicators, multiple causes (MIMIC) model. A MIMIC model appears in figure 3.8. In this figure, we define the quality of a child's home learning environment as a formative construct. Our five exogenous formative indicators are a fairly rich set of indicators: we say that mother's education, household income, number of siblings, parental support, and parental monitoring all combine to form the child's home learning environment. Each formative indicator makes an independent contribution.

In the previous section, we fixed the error variance for our latent variable at 0.0, but here you see that we want to estimate it. Why? Simply because there is more to the quality of a child's home learning environment than our five indicators. We have not included anything about the father's education or his involvement. Because we lack information on these variables, we allow there to be error in our latent variable. We are acknowledging that more goes into the quality of a child's home learning environment than our five formative indicators.

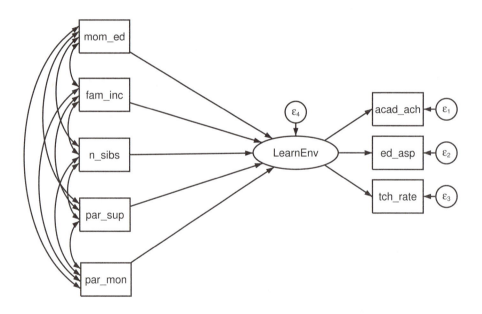

Figure 3.8: MIMIC model

We achieve identification of this formative construct by having some reflective variables flow from the latent variable. In this example, we say that home learning environment leads to (causes) academic achievement, level of educational aspirations, and level of the teacher's ratings. In effect, we have combined five formative indicators with three reflective indicators.

There are just two differences in the `sem` command we would use for this model. First, we do not need to fix one of the formative indicators at 1.0, because we already have a reference indicator (academic achievement). Second, we do not need an option of `var(e.LearnEnv@0)`, because we are able to allow for some error variance in our latent variable.

Box 3.5. Heywood case—negative error variances

In some models, the "best" maximum likelihood estimate is an inadmissible solution. The Heywood case illustrates this problem, where the "best" estimate has one or more negative error variances. An error variance can be very small and may even be zero if your measurement is perfect; however, you cannot have a negative variance. Some SEM programs estimate these error variances as negative values and hope you notice the problem. Stata's `sem` command will not estimate a negative error variance and will instead fail to converge. The obvious disadvantage is that you do not get immediate results; this feature of Stata is advantageous, though, in that it forces you to confront the possibility of a Heywood case.

At the top of the results, Stata reports the log-likelihood function for each iteration. Where there is a problem, Stata reports "not concave", which means that the log-likelihood function is essentially flat at a particular location. This is not a problem if it happens in some iterations, but when there is a long list of iterations that are not concave, Stata may never converge on a solution. If this happens, you can examine the results for an iteration that is not concave to see if there is an error variance that is very close to 0. To get these intermediate results, you add `iterate(#)` as an option, where # is the number of an iteration that is not concave. This option forces Stata to print out the results it has at that iteration. Here are the first seven iterations from a result reported in this chapter:

```
Iteration 0: log likelihood = -15365.508 (not concave)
Iteration 1: log likelihood = -15330.166 (not concave)
Iteration 2: log likelihood = -15320.492 (not concave)
Iteration 3: log likelihood = -15304.647
Iteration 4: log likelihood = -15295.115 (not concave)
Iteration 5: log likelihood = -15284.76 (not concave)
Iteration 6: log likelihood = -15279.147
Iteration 7: log likelihood = -15271.209
```

The model eventually converged on the 15th iteration. If you have what seems like an endless list of not concave iterations, you might find an error variance that is very close to 0, which may be the reason your model is not converging. If you do not see any other problem in the way your model is specified, you might try fixing the offending error variance at 0 or some other very small value.

The MIMIC model makes a lot of sense in many situations. A MIMIC model achieves identification, which is a necessary thing but is not sufficient for a useful model. For a MIMIC model to be useful, you need to theoretically justify what you are doing. It is also very helpful for the reflective indicators (academic achievement, level of educational aspirations, and level of the teacher's ratings) to be moderately correlated.

We could consider many other extensions. A simple one might be to add an additional direct path from mother's education to academic achievement. This way, mother's education directly influences the academic achievement of her children but also indirectly influences it by creating a strong home learning environment. Brown (2006) is a good place to read about MIMIC models and their extensions.

3.6   Exercises

1. A modification index for the covariance between a pair of error terms is 40.00. What does this mean? When should you revise your model to allow for this covariance?

2. What are the differences between the four levels of invariance?

3. Draw a simple model and explain how you would test whether two parameter estimates were equal. Why is testing the equality important?

4. Give an example of a composite latent variable.

   a. Why does the arrow go in the opposite direction for this type of latent variable?

   b. What assumptions do we make to identify a composite latent variable?

5. Make up an example of a MIMIC model that would be useful in your substantive area. Use the SEM Builder to draw this figure.

6. Does a patient's confidence in his or her primary care physician influence the patient's compliance? Does a patient's compliance influence his or her confidence? You have two measures of confidence: a) a scale of the patient's rating of the physician's medical expertise and b) a scale of the patient's rating of how caring the physician is. You also have two measures of compliance: a) compliance about prescriptions and b) compliance to behavioral instructions given by the physician. You have measured all four variables at two time points, and your fit model looks like this:

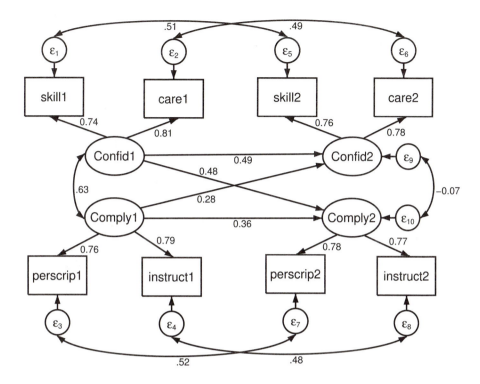

Your data are in the `compliance.dta` dataset, and they are hypothetical.

a. Use the SEM Builder to draw this model as it appears above.

b. Use the SEM Builder to fit this model.

c. Why were the error terms correlated? Justify your answer.

d. How stable is confidence? Compliance?

e. Why might confidence cause compliance? Why might compliance cause confidence? Justify that the influence goes both ways.

f. Is the effect of confidence on compliance greater or lesser than the effect of compliance on confidence?

g. Produce a drawing with results, like the one above, but add asterisks to indicate significance and add other information about the fit and explained variance. If you need help relocating numbers and changing number formats, review appendix A.

4 Latent growth curves

4.1 Discovering growth curves

In chapters 2 and 3, we were predicting a person's score on an observed or latent variable. We identified the variables that predicted who would score higher or lower on an endogenous variable. For example, we could use a path model (chapter 2) or a full structural equation model (chapter 3) to predict who would consume a very high or very low amount of alcohol. In chapter 3, we used an example involving alienation. We predicted a person's level of alienation in 1971 from his or her level of alienation in 1967 and socioeconomic status in 1966. In this chapter, we go over a fundamentally different way of examining data. Instead of predicting a person's score on a variable, we want to

- identify the trajectory, and

- predict who has the more positive or more negative trajectory.

We know, for example, that alcohol consumption goes up dramatically from ages 18 to 23. This is a positive overall trajectory. However, different individuals have different trajectories. In spite of the overall trend, some will maintain a very low level of alcohol consumption, while others will have a much more dramatic increase than the overall trend. Of course, some will be very close to the overall trend. Finding out who has these different trajectories is as important as identifying the overall trajectory in the first place.

The overall trajectory is referred to as a fixed effect; it is what would happen if everybody had the same trajectory. For many purposes, just identifying this fixed effect is sufficient. The overall trajectory has clear implications for broad policy decisions, because it identifies where there is a problem, whether the problem is increased alcohol consumption, increased unsafe sex, or increased body mass index (BMI). Identifying and estimating the fixed effect is our first task in this chapter.

The difference in an individual's trajectory from the overall trajectory is referred to as a random effect. The `sem` command does not estimate a different trajectory for each individual, but it does estimate the variance of the random effects. There will always be some individual differences in trajectories, but when these results produce a significant variance, we say that there are significant random effects in the trajectory.

Actually, there are two possible random effects for a growth trajectory. The first is the intercept or initial level. Some people start higher or lower on alcohol consumption, for example. A latent variable can be used to represent the random intercept. This latent variable may also be called the latent intercept growth factor. When there is substantial variance in the intercepts of different people, we may want to look for covariates that help explain this variance. Do men or women have a higher intercept? Do more religious people have a lower intercept? Do people from the "Bible Belt" states have a lower intercept?

The second possible random effect is the slope or rate of change. Why do some people experience a dramatic increase in BMI during their 20s while others maintain a constant BMI and still others actually lower their BMI in their 20s? Perhaps education plays a role, with more educated people having a less dramatic increase in BMI. Perhaps gender is relevant. Could race/ethnicity be important? Surely a person's exercise routine is relevant to both the intercept and the slope of his or her BMI trajectory. Our predictors are now explaining trajectories rather than scores at just one time point. Our latent variables represent an intercept and a slope.

4.2   A simple growth curve model

In a simple linear growth curve, we need to identify the intercept to know where the growth curve starts, and then we need to identify the slope to know the rate of increase or decrease that occurs for each unit change in time. Both the intercept and slope are treated as latent variables that we need to identify using structural equation modeling. Good names for these are latent intercept growth factor and latent slope growth factor, but the shortened versions, intercept and slope, work fine.

With a single intercept and slope, we are usually describing a linear growth curve. We use the term "curve" loosely, recognizing that it may be a straight line, a quadratic curve, or some other shape. In our simple example, we are asserting that, over time, the trajectory moves up or down in a straight line. This may make sense for many variables, but sometimes a straight line is insufficient to accurately describe the overall trajectory. Alcohol consumption might have a fairly linear positive trajectory between the ages of 18 and 23. By contrast, alcohol consumption might have a fairly linear negative trajectory from ages 23 to 30. If we wanted to describe the trajectory of alcohol consumption from ages 18 to 30, however, we would probably want a quadratic slope to allow the trajectory for alcohol consumption to increase at first and then decrease sometime after age 23.

Let us begin with the simplest example, a linear growth curve like the one in figure 4.1.

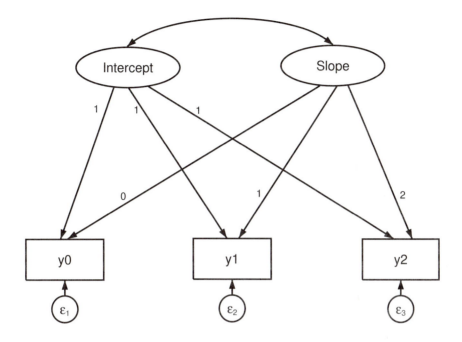

Figure 4.1. A linear growth curve

The variables y0, y1, and y2 are observed variables. They indicate how people score on whatever we are measuring. The y0 might be the person's score at the baseline; for example, if these observed variables represented alcohol consumption at ages 18, 20, and 22, then y0 would be the person's score on alcohol consumption at age 18. If these observed variables were testing the effects of an intervention, the y0 might be the score before the intervention commenced; y1 would be the person's score at wave 2, which could be the score at the end of the intervention; and y2 would be the person's score at wave 3, maybe at a follow-up measurement one month after the intervention was completed.

Although these observed variables refer to times/waves one, two, and three, we relabeled them as 0, 1, and 2. This is so that the first variable, y0, is the person's score at the start, reflecting his or her intercept. The three observed scores are all the measurements we have for each individual. The other variables in the model appear inside ovals; these are latent variables representing the error terms plus the intercept and slope.

We assume there will be some error in our measurement at each wave, so we have latent error terms—ϵ_1, ϵ_2, and ϵ_3—to allow for this error. Similarly, we allow for some random variation from one person to the next in the person's latent intercept. The difference from the overall intercept corresponds to the random effect in his or her latent intercept, because some people will have a much higher initial level than others. We do not actually estimate each person's difference from the overall intercept, but we estimate the variance of these random effects.

Similarly, we allow for some variation from one person to the next in the latent slope growth factor, and we call that difference from the overall slope the random effect of the slope. Again, we estimate the variance of this random effect. Notice in figure 4.1 that we have even allowed the latent intercept growth factor and the latent slope growth factor to be correlated. This is the curved line with an arrow at both ends that connects the intercept and slope growth factors.

Consider a study of what happens to people's BMI during their 20s. We might expect people who have a low BMI at age 20 (their individual intercept) to have a low growth rate (their individual slope). By contrast, we might expect people who have a high BMI at age 20 to have a larger than average rate of growth in their BMI. If this were true, then the intercept and slope growth factors would be positively correlated—the higher the initial BMI, the greater the rate of growth in BMI. This would indicate a cumulative disadvantage and, if true, would be an important addition to our understanding of obesity.

To identify the latent intercept and latent slope, remember your basic regression equation: the intercept is the constant to which you add or subtract the effect of change in your predictor. Most statistical packages, including Stata, refer to the intercept as the coefficient on the constant. Figure 4.1 represents this constant by assigning a fixed value of 1.0 to the path from the latent intercept to each of the observed variables.

From your basic regression training, you will remember that the slope is the change in the outcome for each unit change in your predictor. In figure 4.1, we identify these values at 0, 1, and 2. (It may look strange to have a path with a 0 on it, but showing it this way may clarify what happens.) If y is measured at three equal-interval time points, then we can say that the first time point is 0, the second is 1, and the third is 2; the values we choose depend completely on how time is measured. If you are using a person's score at age 18, again at age 20, and finally at age 22, you might use 0, 2, and 4 because each wave is two time units (years) of time apart. If you use 0, 1, and 2, you need to remember that one unit represents 2 years when you interpret your results.

In an intervention, let us say that you have a baseline, a score at completion of the intervention (12 weeks later), and a follow-up score at 1 year from the baseline. You could use 0, 12, and 52, where each unit of change in time represents 1 week. Alternatively, you might use 0, 3, and 13, where each unit of change in time represents 4 weeks. What is important is that the numbers begin with 0 and reflect how much time expires between the measurements. There are specialized applications where the intercept is defined at the midpoint or even at the endpoint, such as time to death.

4.3   Identifying a growth curve model

Identification is crucial but fairly complicated. Figure 4.1 shows that we have three time points for measuring our observed variables: y0, y1, and y2. We have $\{k(k+1)\}/2$ variances and covariances, where k is the number of observed variables in our model. For figure 4.1, $k = 3$, so we have $\{3 \times (3+1)\}/2 = 6$ variances and covariances.

$$\begin{bmatrix} \sigma_1^2 & & \\ \sigma_{21} & \sigma_2^2 & \\ \sigma_{31} & \sigma_{32} & \sigma_3^2 \end{bmatrix}$$

We also know the means of the observed variables. We thus have three variances, three covariance, and three means. It is possible to write an equation for each of these nine known values.

How many unknowns are we estimating? We estimate three variances for the error terms, along with the two variances and the covariance for the intercept and slope. In addition, we estimate the means of the intercept and slope, which represent the fixed effects. These represent the fixed effects. We also have the covariances of the intercept and the slope. Thus we have nine knowns $(\sigma_1^2, \sigma_2^2, \sigma_3^2, \sigma_{21}, \sigma_{31}, \sigma_{32}, \mu_1, \mu_2, \mu_3)$ with which to estimate the eight unknown parameters (var(e.y0), var(e.y1), var(e.y2), mean(Intercept), mean(Slope), var(Intercept), var(Slope), and cov(Intercept, Slope)) resulting in 1 degree of freedom to test our model. With three time points, we can fit a linear latent growth curve and test its goodness of fit, although the test with just 1 degree of freedom is a minimal test.

What would happen if we had four time points? Then we would have $\{4 \times (4+1)\}/2 = 10$ elements of the covariance matrix (four variances and six covariances) along with four means. We would still have the same parameters we were estimating plus one additional error variance, and so we would have $14 - 9 = 4$ degrees of freedom. Having one more time point would provide a much more rigorous test of our model.

4.3.1   An intuitive idea of identification

Time points added to a model not only add degrees of freedom but also provide more information for testing the model. With just two time points, there is no information to test such a relationship because two points determines only one line; for example, see figure 4.2. With three time points, there is a bit more information to test our linear model, and with four time points, there is even more information for our test. The take-away point to remember is that you need at least three time points for a linear latent growth curve, though four or more time points is best.

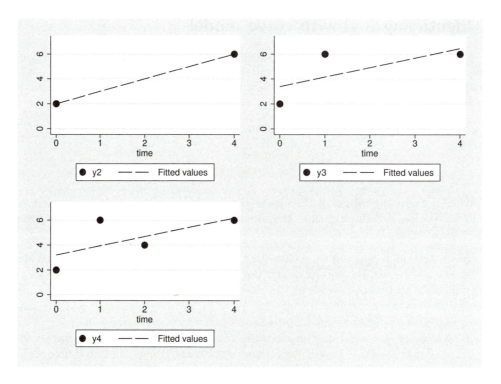

Figure 4.2. A linear relationship with two, three, and four data points

4.3.2 Identifying a quadratic growth curve

With four time points, we can fit a simple curve using a quadratic. Figure 4.3 shows how we would draw such a model. The latent intercept growth factor with loadings of 1 for each wave and the latent linear slope growth factor with loadings of 0, 1, 2, and 3 are unchanged. What is new is the latent quadratic slope growth factor. You may remember that in ordinary regression, you used X to represent the linear predictor ($\widehat{Y} = B + B_1 X$) and X^2 to represent the quadratic ($\widehat{Y} = B + B_1 X + B_2 X^2$). We follow the same logic here. Notice in figure 4.3 that the loadings for the latent quadratic slope are just the square of the loadings for the latent linear slope; thus 0 stays at 0, 1 stays at 1, 2 becomes 4, and 3 becomes 9.

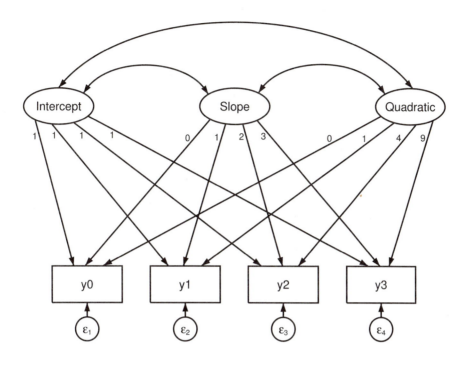

Figure 4.3. Growth curve with linear and quadratic components

The data associated with figure 4.3 have 14 underlying observed moments: $\{4 \times (4 + 1)\}/2 = 10$ variances and covariances plus 4 means. We now have 13 parameters to estimate: variances of 4 error terms (e_1–e_4), 3 means (intercept, linear slope, quadratic slope), 3 variances (intercept, slope, quadratic), and 3 covariances (intercept with slope, intercept with quadratic, slope with quadratic). Thus with 4 time points, we have an identified latent growth curve including both linear and quadratic components with $14 - 13 = 1$ degree of freedom.

With Stata's sem command, we can fit linear and quadratic growth curves, and test to see whether the quadratic is necessary. We should test for a quadratic component whenever we believe the growth trajectory follows a simple curve. With more time points, it is possible to add higher dimensions, but these are often difficult to interpret.

4.4 An example of a linear latent growth curve

We have data from the National Longitudinal Survey of Youth, 1997. This is a national survey of youth who were 12–16 years old in 1997. They have been interviewed each year since then as part of a major study of the transition to adulthood in terms of topics such as employment and education. With a subset of these data, consisting of those who were 16 years old in 1997, we will fit a latent growth curve with the data for the years 2001–2009. We have self-reported information on their height and weight.[1]

In 2001, our subset of participants were 20 years old, and in 2009 they were 28. We are interested in what happens to BMIs of people during their 20s. BMI is calculated as $(\text{weight}/\text{height}^2) \times 703$, where weight is measured in pounds and height is measured in inches. The data are in the public domain, and the complete dataset can be downloaded at http://www.nlsinfo.org/investigator/pages/login.jsp. We are just using a few variables and a subset of years for our example in this chapter; I have also arbitrarily deleted some observations where people reported weighing under 50 pounds or being under 4 feet, 2 inches tall. The dataset we are using is `bmiworking.dta`.

4.4.1   A latent growth curve model for BMI

First, we should draw a model of the growth curve, which is shown in figure 4.4. I have made a couple of stylistic simplifications compared with figures 4.1 and 4.3. I left the loadings out of the figure because it would be too cluttered with them in. This is not a problem with a linear latent growth curve, because the loadings of the intercept to BMI each year is always fixed at 1.0, and the loadings from the slope to BMI each year are 0, 1, 2, 3, 4, 5, 6, 7, and 8. Remember that the curved line with an arrow on each end represents the covariance of the intercept with the slope. As before, the first wave, BMI in 2001, is wave 0.

---

1. Self-reports of height and weight can have serious measurement error, and this error can be biased. Men might over-report their height and under-report their weight if they feel they are overweight. Men who have a very low weight might overestimate their weight. Women may also give biased reports. Because these data are being used as an example, we will not worry about these possible biases; but we should acknowledge that direct observer measurement of height and weight would give us far greater confidence in our findings.

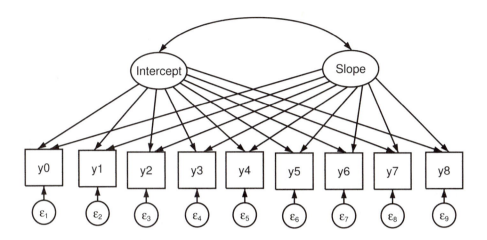

Figure 4.4. Linear latent growth curve for BMI between 20 and 28 years of age

4.4.2 Graphic representation of individual trajectories (optional)

Before jumping into fitting the latent growth curve of BMI from age 20 to age 28, it is helpful to look at a subsample of a few individual trajectories. To create a figure showing individual trajectories requires you to run Stata code that does not use the `sem` command. This is not a necessary step. However, it is nice to see the overall linear trajectory and also the year-by-year change for a small sample of observations—say, 20 to 50 people. The results of these efforts appear in figure 4.7 at the end of this section.

First, we open the dataset and keep just the `bmi01`–`bmi09` observed variables along with the `id` variable.

```
. use http://www.stata-press.com/data/dsemusr/bmiworking.dta
. sort id
. keep id bmi01-bmi09
```

The data are in what is called a wide format. This means we have one row of data for each person. Look at the tiny part of our data shown in figure 4.5. The person whose `id` is 14 has missing values for all nine BMI scores, `bmi01`–`bmi09`. The next person has a missing score on all the variables except for `bmi03`. You may notice that we have no variable `bmi04` for when the person was 23; this variable was left out so that you can see how to adjust for a missing wave in the `sem` command.

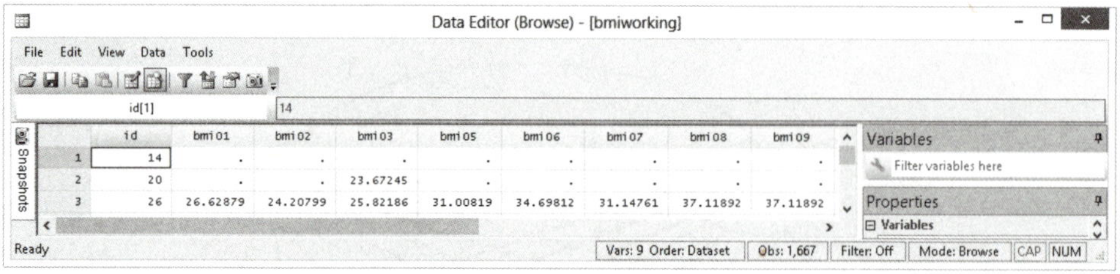

Figure 4.5. Subset of our data in wide format

To construct our graph, we need to reshape the data into what is called a long format. I will not go into the details of this extremely powerful command other than to illustrate how we use it. The command we run is

```
. reshape long bmi0, i(id) j(year)
```

The **reshape long** part of the command tells Stata that we are going to go from a wide way of arranging the data to a long way. The **bmi0** is the stub for each of our variables, **bmi01**, **bmi02**, ..., **bmi09**. The stub includes everything except the numerical value that distinguishes each wave; we include the 0 in the stub because each wave has a 0 in its name. If our variables were **hlth1**, **hlth2**, ..., **hlth10**, then our stub would be **hlth**, and the first part of our command would be **reshape long hlth**.

After the comma, we have **i(id) j(year)**. The **i(id)** is needed to identify each person uniquely. We have labeled **id** as the identifier because this variable contains the unique identification number. The **j(year)** is what makes this **reshape** command so ingenious. It will create a new variable called **year** that will take on values of 1, 2, 3, 5, 6, 7, 8, 9 (notice the missing 4). The number at the end of each variable name is used to create this variable, and we have no variable **bmi04** in our dataset. Figure 4.6 illustrates how the first three people (IDs 14, 20, and 26) appear in the long format.

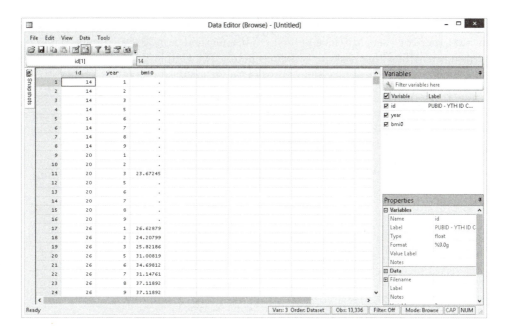

Figure 4.6. Subset of our data in long format

Person 14 had no data on any of the BMI scores. Person 20 only had a score for the third year of data, 2003, when the person was 22 years old. Person 26 had 8 years of BMI data. There is just one BMI variable, `bmi0`, because we now have our `year` variable to distinguish the waves. Check out person 26's BMI in the last wave, wave 9. Person 26 had a BMI of 37.12. By contrast, when person 26 was just 21 years old at wave 2, the BMI was 24.21. A BMI over 25 is considered overweight, so person 26 went from a normal weight to being morbidly obese during his or her 20s.

Before we create our graph, we need to do a linear regression to see the overall trajectory. We first label the values on the `year` variable. The last two commands run a regression and then predict the outcome score based on the regression. We include the option `cluster(id)` to adjust for the clustering of the eight observations within each of the individuals; we are treating each individual as a cluster of eight repeated scores. This option only affects the standard errors and hence is not necessary to create the graph.

```
. label define age 1 "20" 2 "21" 3 "22" 4 "23" 5 "24" 6 "25" 7 "26" 8 "27" 9 "28"
. label values year age
. regress bmi0 year, cluster(id)
  (output omitted )
. predict yhat
(option xb assumed; fitted values)
```

There is a better way to do this by using the `mixed` command along with `margins` and `marginsplot` (see Mitchell [2012]). This method requires an understanding of multilevel modeling that I do not assume here. The simple `regress` command gives you a rough visual idea of how the BMI changes for a subsample of our observations.

Now we are ready to create our graph. We can use the **Graphics > Twoway graph (scatter, line, etc.)** dialog box to prepare our command. We will create one graph with just the linear predicted value and then a separate graph for 10 randomly selected people. We then overlay these graphs into a single figure.

The complete graph command is quite lengthy, but the syntax is simply repeated for each line:

```
. twoway (lfit yhat year, lwidth(thick))                        ///
  (line bmi0 year if id == 14,                                  ///
    lcolor(black) lwidth(medthin) lpattern(longdash))           ///
  (line bmi0 year if id == 137,                                 ///
    lcolor(black) lwidth(medthin) lpattern(longdash))           ///
  (line bmi0 year if id == 153,                                 ///
    lcolor(black) lwidth(medthin) lpattern(longdash))           ///
  (line bmi0 year if id == 212,                                 ///
    lcolor(black) lwidth(medthin) lpattern(longdash))           ///
  (line bmi0 year if id == 216,                                 ///
     lcolor(black) lwidth(medthin) lpattern(longdash))          ///
  (line bmi0 year if id == 260,                                 ///
     lcolor(black) lwidth(medthin) lpattern(longdash))          ///
  (line bmi0 year if id == 287,                                 ///
     lcolor(black) lwidth(medthin) lpattern(longdash))          ///
  (line bmi0 year if id == 1488,                                ///
     lcolor(black) lwidth(medthin) lpattern(longdash))          ///
  (line bmi0 year if id == 2403,                                ///
     lcolor(black) lwidth(medthin) lpattern(longdash))          ///
  (line bmi0 year if id == 8924,                                ///
     lcolor(black) lwidth(medthin) lpattern(longdash)),         ///
     ytitle(Body mass index (BMI)) xtitle(Age)                  ///
     xlabel(1(1)9, valuelabel) legend(off)
```

The command above produces the graph in figure 4.7. The thick line in the figure shows the overall linear trajectory. The dashed lines show the actual changes from year to year for our 10 randomly selected people. The overall trajectory is positive, with BMI increasing from year to year. When we look at the 10 individual line graphs, we see some people with missing values, some with no missing values, some who bounce up and down more than others; however, there is an overall upward trend in BMI during our respondents' 20s.

The overall intercept looks like it is about 26, but notice the substantial variance between individuals: some people start with a BMI of over 30, and some start with a BMI of around 20. When there is a lot of variance in the intercept, we need to use what is called a random intercept model. We need to estimate the intercept but also the amount of variance around the intercept. We have a similar result with the slope: some trajectories are positive with fairly steep inclines, while some have a slope near 0 or even slightly negative.

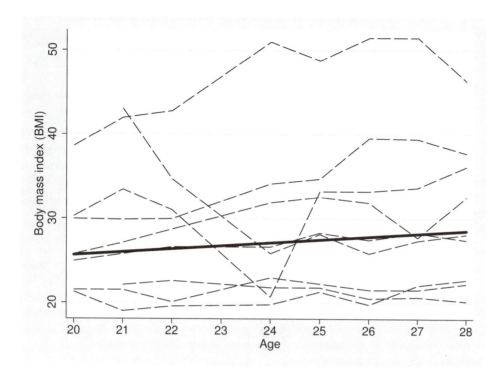

Figure 4.7. Growth patterns for a sample of people

When there is a lot of variance—that is, significant random effects in both the intercept and slope—we have what is often called a random coefficients or random slopes model. When this happens, in addition to estimating the overall trajectory and the variation of the random effects, we may also want to add covariates that explain the random intercept and slope. That is, if we have covariates that explain why some people are initially higher on BMI than others and why some people have a rapid increase in BMI while others have little change or a decline in BMI, we can add these covariates to the model. If we were to add covariates that do not change with time to the random coefficients model, the latent intercept growth factor and latent slope growth factor would become dependent variables—we would now allow for differences in overall trajectories based on these covariates in addition to allowing for random variation in the slope and intercept.

4.4.3 Intraclass correlation (ICC) (optional)

The bold regression line we plotted in figure 4.7 is not ideal because it ignores the interdependence of the repeated measures on the same people. People who have a high score one year are more likely to have high scores other years. We would expect some degree of consistency, and this means that our observations are not going to be statistically independent as is assumed for ordinary least-squares (OLS) regression. For example, we expect that your score when you were 20 years old will be related to your score when you were 21 years old.

Because our data are in the long format, we can quickly check for this lack of statistical independence by estimating the intraclass correlation (ICC), also called ρ_I ("rho"). The higher the ICC, the greater the dependence and the more inappropriate the OLS regression would be. Values close to 0.0—say, less than 0.05—are sometimes used to justify using OLS regression. However, when a study is designed to have repeated measures, that is, multiple scores nested in each individual, then there is a good reason to go beyond depending on OLS, regardless of the size of the ICC. OLS offers no theoretical advantage, even when the ICC is small. However, if the ICC is very close to 0, you may not have convergence when fitting the model that includes a random intercept.

With the data in long format, a way to estimate the ICC is to use the `mixed` command:

```
. mixed bmi0 || id:

Performing EM optimization:

Performing gradient-based optimization:

Iteration 0:   log likelihood = -27467.968
Iteration 1:   log likelihood = -27467.968

Computing standard errors:

Mixed-effects ML regression              Number of obs      =      10365
Group variable: id                       Number of groups   =       1581

                                         Obs per group: min =          1
                                                        avg =        6.6
                                                        max =          8

                                         Wald chi2(0)       =          .
Log likelihood = -27467.968              Prob > chi2        =          .
```

bmi0	Coef.	Std. Err.	z	P>|z|	[95% Conf. Interval]	
_cons	27.07653	.146103	185.32	0.000	26.79018	27.36289

Random-effects Parameters	Estimate	Std. Err.	[95% Conf. Interval]	
id: Identity				
var(_cons)	32.42487	1.198465	30.15899	34.861
var(Residual)	7.013245	.1057932	6.808929	7.223691

```
LR test vs. linear regression: chibar2(01) = 12611.35 Prob >= chibar2 = 0.0000
. estat icc

Intraclass correlation
```

Level	ICC	Std. Err.	[95% Conf. Interval]	
id	.8221709	.0058677	.8103776	.8333816

The `mixed` command is shown with the outcome variable, `bmi0`, and no predictors. After the pair of vertical bars, `||`, is the name of the identification variable that specifies how the data are clustered. In this example, each individual is a cluster and the identification variable for individuals is `id` (it is necessary to include the colon after the identification variable). The `estat icc` command gives us the ICC based on the results of the `mixed` model.

The ICC of 0.82 indicates a very high degree of dependence in BMI scores across the waves. The variance between individuals is shown in the `mixed` results as `var(_cons)` and is 32.42. This represents the differences between people. The variance within people (across the repeated measures) is reported as the `var(Residual)` and is 7.01. This represents how individuals vary from year to year. We can square these to get variances and then estimate the ICC.

$$ICC = \frac{\text{Between variance}}{\text{Between variance} + \text{Within variance}}$$
$$= \frac{32.42}{32.42 + 7.01}$$
$$= 0.82$$

Stata has alternative ways of estimating the ICC; for example, you could use the command `icc bmi0 id`. This command applies when you have a balanced design (no missing values). There are usually at least some missing values, making our growth curve an unbalanced design. The `icc` command uses casewise deletion, dropping any individual that has any missing values. In our example, `icc bmi0 id` yields ICC $= 0.83$ for the 781 observations that have a value at all eight waves of data. We want to use all of our available data for the 1,581 observations and so rely on `mixed bmi0 || id:` followed by `estat icc`.

4.4.4   Fitting a latent growth curve

We can fit a basic linear growth curve either using listwise deletion or using all available data. We just saw from the `mixed` results that the average number of BMI reports was 6.6 waves for our 1,581 individuals. It is extremely common to have substantial missing values like this in a longitudinal study. People are unavailable for one or more years because of any number of reasons (out of town during the interview period, incarcerated, in military service, refuse to answer, ill at the time of the interview).

We can either use all the information we have for each individual, missing values and all, or we can just keep the individuals who have complete data. Only 781 people have complete data that allows us to calculate BMI for all eight years. In all likelihood, limiting our sample to just these 781 individuals and thereby throwing out more than 50% of our subset will introduce more bias than using all the available information we have on the 1,581 individuals. By using all available data, we assume the missing values are missing at random and that variables have a multivariate normal distribution.

We will use the maximum likelihood estimator. The default maximum likelihood estimator uses listwise deletion and has $N = 781$. The full-information maximum likelihood estimator has $N = 1,581$. We will compare them later. For now, we will see the commands for the default maximum likelihood estimator with listwise deletion. We apply this on the original dataset in wide format by reopening the dataset. We will soon see a problem with this command:

```
. use http://www.stata-press.com/data/dsemusr/bmiworking.dta, clear
. sem (Intercept@1 Slope@0 -> bmi01)      ///
      (Intercept@1 Slope@1 -> bmi02)      ///
      (Intercept@1 Slope@2 -> bmi03)      ///
      (Intercept@1 Slope@4 -> bmi05)      ///
      (Intercept@1 Slope@5 -> bmi06)      ///
      (Intercept@1 Slope@6 -> bmi07)      ///
      (Intercept@1 Slope@7 -> bmi08)      ///
      (Intercept@1 Slope@8 -> bmi09)
```

This command is lengthy by Stata standards, but it is a fairly simple way to describe our model in figure 4.4. This simple code, in combination with Stata's default assumptions for sem, fully describes the figure for a latent linear growth curve. We have an intercept fixed at 1 for each year. For the slopes, we use the coding described earlier, that is, 0, 1, 2, 4, 5, 6, 7, 8. The first year has a slope fixed at 0.0, the second year at 1.0, and the third year at 2.0. We do not include bmi04 because we do not have BMI data for that year. We skip this variable and fix the slope for bmi05 at 4, for bmi06 at 5, for bmi07 at 6, for bmi08 at 7, and for bmi09 at 8. The direction of the arrow in each line of the command is critical. As shown in figure 4.4, our latent variables, Intercept and Slope, are shown as causing the BMI scores. Your BMI for any given year depends on where you started and your trajectory.

Let us look at partial results and see what the problem is with the way we wrote this command:

```
(886 observations with missing values excluded)
Endogenous variables
Measurement:  bmi01 bmi02 bmi03 bmi05 bmi06 bmi07 bmi08 bmi09
Exogenous variables
Latent:       Intercept Slope
 (output omitted)
Structural equation model                Number of obs      =        781
Estimation method  = ml
Log likelihood     = -15043.332
 (output omitted)
```

	Coef.	OIM Std. Err.	z	P>|z|	[95% Conf. Interval]	
Measurement						
bmi01 <-						
Intercept	1	(constrained)				
_cons	25.3212	.198602	127.50	0.000	24.93194	25.71045
bmi02 <-						
Intercept	1	(constrained)				
Slope	1	(constrained)				
_cons	25.9587	.2020412	128.48	0.000	25.5627	26.35469
bmi03 <-						
Intercept	1	(constrained)				
Slope	2	(constrained)				
_cons	26.36276	.2072692	127.19	0.000	25.95652	26.769
bmi05 <-						
Intercept	1	(constrained)				
Slope	4	(constrained)				
_cons	27.0444	.2177605	124.19	0.000	26.6176	27.47121
bmi06 <-						
Intercept	1	(constrained)				
Slope	5	(constrained)				
_cons	27.55184	.2320204	118.75	0.000	27.09709	28.00659
bmi07 <-						
Intercept	1	(constrained)				
Slope	6	(constrained)				
_cons	27.73508	.2241188	123.75	0.000	27.29582	28.17435
bmi08 <-						
Intercept	1	(constrained)				
Slope	7	(constrained)				
_cons	28.08077	.2298791	122.15	0.000	27.63021	28.53132

```
       bmi09 <-
      Intercept |          1  (constrained)
          Slope |          8  (constrained)
          _cons |   28.26316   .2409994   117.27   0.000    27.79081    28.73551
     -----------+----------------------------------------------------------------
    var(e.bmi01) |   2.244796   .2062215                      1.874908    2.687657
    var(e.bmi02) |   2.963643   .2018882                      2.593228    3.386968
    var(e.bmi03) |   3.895976   .2320473                      3.466714     4.37839
    var(e.bmi05) |   4.755829    .265343                      4.263193    5.305392
    var(e.bmi06) |   7.881205   .4242931                      7.091976    8.758263
    var(e.bmi07) |   2.800865   .1808624                      2.467897    3.178758
    var(e.bmi08) |     2.1962   .1681625                      1.890147    2.551809
    var(e.bmi09) |   3.257031   .2439416                      2.812351    3.772022
   var(Intercept) |      28.56   1.506151                     25.75544    31.66996
       var(Slope) |    .190818   .0122179                       .168313    .2163321
     -----------+----------------------------------------------------------------
   cov(Interc-t,
          Slope) |   .0832273    .095453     0.87   0.383     -.103857    .2703117
```

LR test of model vs. saturated: chi2(25)   =   323.78, Prob > chi2 = 0.0000

We are first informed that we have excluded 866 observations with missing values (this includes those who had no data for any year as well as those who had data for fewer than the full 8 years). Our sample size is now 781. For each year, we get the fixed values for the intercept, slope, and an estimated coefficient on the constant. The coefficients for each of the intercepts are 1.0 because these were fixed (reported as constrained in the output). The coefficients for the slopes are also fixed values: 0, 1, 2, 4, 5, 6, 7, 8. The coefficient on the constant for each wave is nothing other than the mean BMI for our sample of 781 people for that year.

Have you guessed what is missing? The mean of the latent intercept growth factor and the mean of the latent slope growth factor do not appear in our results. We are missing exactly what we are trying to estimate.

We must have a mean for the intercept and a mean for the slope because those two mean values determine the overall trajectory. We have to ask for these values by adding the options `noconstant` and `means(Intercept Slope)`. We will also include an option to give us a full-information maximum likelihood estimation: `method(mlmv)`, where `mlmv` stands for maximum likelihood estimation with missing values. The added options appear in the last line of the command.

```
. use http://www.stata-press.com/data/dsemus/bmiworking.dta, clear
. sem (Intercept@1 Slope@0 -> bmi01)
>     (Intercept@1 Slope@1 -> bmi02)
>     (Intercept@1 Slope@2 -> bmi03)
>     (Intercept@1 Slope@4 -> bmi05)
>     (Intercept@1 Slope@5 -> bmi06)
>     (Intercept@1 Slope@6 -> bmi07)
>     (Intercept@1 Slope@7 -> bmi08)
>     (Intercept@1 Slope@8 -> bmi09),
>     method(mlmv) noconstant means(Intercept Slope)
(86 all-missing observations excluded)
```
   (output omitted)

```
Structural equation model                    Number of obs     =       1581
Estimation method  = mlmv
Log likelihood     = -26067.646
```
   (output omitted)

		OIM				
	Coef.	Std. Err.	z	P>|z|	[95% Conf. Interval]	
Measurement						
bmi01 <-						
Intercept	1	(constrained)				
_cons	0	(constrained)				
bmi02 <-						
Intercept	1	(constrained)				
Slope	1	(constrained)				
_cons	0	(constrained)				
bmi03 <-						
Intercept	1	(constrained)				
Slope	2	(constrained)				
_cons	0	(constrained)				
bmi05 <-						
Intercept	1	(constrained)				
Slope	4	(constrained)				
_cons	0	(constrained)				
bmi06 <-						
Intercept	1	(constrained)				
Slope	5	(constrained)				
_cons	0	(constrained)				
bmi07 <-						
Intercept	1	(constrained)				
Slope	6	(constrained)				
_cons	0	(constrained)				
bmi08 <-						
Intercept	1	(constrained)				
Slope	7	(constrained)				
_cons	0	(constrained)				

bmi09 <-						
Intercept	1	(constrained)				
Slope	8	(constrained)				
_cons	0	(constrained)				
mean(Interc~t)	25.62992	.139267	184.03	0.000	25.35696	25.90288
mean(Slope)	.3511752	.0134078	26.19	0.000	.3248964	.377454
var(e.bmi01)	2.439363	.1938763			2.08749	2.850549
var(e.bmi02)	4.190178	.2182022			3.783611	4.640433
var(e.bmi03)	4.76602	.2335162			4.329626	5.2464
var(e.bmi05)	5.773539	.2593655			5.286929	6.304937
var(e.bmi06)	8.212432	.3516456			7.551347	8.931391
var(e.bmi07)	3.797888	.1904744			3.442326	4.190175
var(e.bmi08)	2.926603	.1689021			2.613597	3.277094
var(e.bmi09)	3.305474	.2090666			2.920091	3.741718
var(Intercept)	28.49238	1.089272			26.43547	30.70933
var(Slope)	.1897726	.0097477			.1715977	.2098726
cov(Interc~t, Slope)	.0726611	.0734359	0.99	0.322	-.0712707	.2165929

LR test of model vs. saturated: chi2(31) =   376.67, Prob > chi2 = 0.0000

We have all 1,581 observations that have a BMI score for at least 1 year. The first several lines in the table of results show the coefficients on the Intercept and Slope where we have fixed values. Stata says these values are constrained, meaning they are forced to be these particular values. Below that are the means: the Intercept mean is 25.63 and the Slope mean is 0.35. This means the expected gain in BMI is 0.35 points each year between the ages of 20 and 28. An increase of 0.35 points may not sound like much, but over 9 years, you can see that it will add up. We could write an equation to represent our growth curve:

$$\mathrm{BMI_{est}} = 25.63 + 0.35 \times \mathrm{Year}$$

At age 20, wave 0, our estimated BMI is $25.63 + 0.35 \times 0 = 25.63$. Eight years later at age 28, wave 9, our estimated BMI is $25.63 + 0.35 \times 8 = 28.43$. This is not happy news for people in their 20s! During their 20s, people go from a BMI slightly above the cutoff of 25.0 for being overweight to a BMI of 28.44, which is clearly overweight (the cutoff for obese is 30.0).

How well does our linear model fit the data? At the bottom of our results, we see one of the most important criterions: LR test of model vs. saturated: chi2(31) = 376.67, Prob > chi2 = 0.0000. A saturated model, by definition, has no degrees of freedom and perfectly reproduces the known moments (variances, covariances, means). Our model, which has 31 degrees of freedom, fails to perfectly reproduce all the known moments. In the strictest sense, our model should be rejected. However, it is often the case that a model can be imperfect based on this chi-squared test and still be very useful. We can use the postestimation command estat gof, stats(all) to get a more comprehensive view of how well our model fits the data.

```
. estat gof, stats(all)
```

Fit statistic	Value	Description
Likelihood ratio		
chi2_ms(31)	376.672	model vs. saturated
p > chi2	0.000	
chi2_bs(28)	15415.481	baseline vs. saturated
p > chi2	0.000	
Population error		
RMSEA	0.084	Root mean squared error of approximation
90% CI, lower bound	0.077	
upper bound	0.092	
pclose	0.000	Probability RMSEA <= 0.05
Information criteria		
AIC	52161.292	Akaike´s information criterion
BIC	52231.047	Bayesian information criterion
Baseline comparison		
CFI	0.978	Comparative fit index
TLI	0.980	Tucker-Lewis index
Size of residuals		
CD	0.996	Coefficient of determination

Note: SRMR is not reported because of missing values.

Our linear growth curve has a likelihood ratio $\chi^2(31) = 376.67$, $p < 0.001$, repeating what we already know. Clearly, a linear trajectory does not perfectly describe what happens to BMI during your 20s. The root mean squared error of approximation (RMSEA) = 0.08, which is larger than we would like; we want the RMSEA to be less than 0.05. On the other hand, the comparative fit index (CFI) = 0.98, which is above the recommended minimum value of 0.95. It looks like the model fits okay as a rough approximation, and some researchers would be satisfied with this as a working model. However, there may be room for improvement.

We can examine the modification indices to see whether they suggest any possible solutions to our marginal model fit.

```
. estat mindices
```

Modification indices

		MI	df	P>MI	EPC	Standard EPC
Measurement						
bmi01 <-						
	Intercept	40.950	1	0.00	-.0166336	-.0159643
	Slope	6.153	1	0.01	-.3763593	-.0294793
	_cons	29.236	1	0.00	-.369401	-.0664195
bmi02 <-						
	Intercept	8.616	1	0.00	.0075277	.0069928
	_cons	5.286	1	0.02	.1541014	.0268184
bmi03 <-						
	Intercept	12.291	1	0.00	.0092085	.0083918
	_cons	8.822	1	0.00	.2045	.0349136
bmi05 <-						
	Slope	5.077	1	0.02	.3151265	.0223037
bmi06 <-						
	Intercept	9.445	1	0.00	.009824	.0080746
	Slope	7.818	1	0.01	.4518062	.0303066
	_cons	7.480	1	0.01	.2286944	.0352147
bmi07 <-						
	Slope	5.586	1	0.02	.2823065	.0194464
bmi08 <-						
	Slope	22.987	1	0.00	.5704814	.0384687
bmi09 <-						
	Intercept	18.730	1	0.00	-.0110376	-.0087725
	Slope	83.091	1	0.00	-1.224257	-.0794095
	_cons	14.574	1	0.00	-.2551494	-.0379908

cov(e.bmi01,e.bmi02)	13.994	1	0.00	.8412867	.2631416
cov(e.bmi01,e.bmi03)	12.755	1	0.00	-.6862712	-.2012703
cov(e.bmi01,e.bmi05)	26.883	1	0.00	-.9014917	-.2402163
cov(e.bmi01,e.bmi06)	19.306	1	0.00	-.8526607	-.1905031
cov(e.bmi01,e.bmi09)	33.539	1	0.00	.8732158	.3075151
cov(e.bmi02,e.bmi03)	40.765	1	0.00	1.174939	.2629184
cov(e.bmi02,e.bmi05)	10.501	1	0.00	-.5927173	-.1205065
cov(e.bmi02,e.bmi07)	11.790	1	0.00	-.5255302	-.1317378
cov(e.bmi02,e.bmi08)	5.630	1	0.02	-.3411756	-.0974271
cov(e.bmi02,e.bmi09)	9.481	1	0.00	.4879905	.1311228
cov(e.bmi03,e.bmi05)	11.655	1	0.00	.6379434	.1216139
cov(e.bmi03,e.bmi07)	5.767	1	0.02	-.3789351	-.0890668
cov(e.bmi05,e.bmi06)	54.251	1	0.00	1.658588	.2408693
cov(e.bmi05,e.bmi07)	25.018	1	0.00	.8325464	.1777937
cov(e.bmi05,e.bmi09)	19.652	1	0.00	-.7834885	-.1793472
cov(e.bmi06,e.bmi07)	28.887	1	0.00	1.029289	.1843021
cov(e.bmi06,e.bmi08)	4.765	1	0.03	-.4059348	-.0828015
cov(e.bmi06,e.bmi09)	25.880	1	0.00	-1.047621	-.2010719
cov(e.bmi07,e.bmi08)	11.256	1	0.00	.5198754	.155936
cov(e.bmi07,e.bmi09)	32.684	1	0.00	-1.001137	-.2825566
cov(e.bmi08,e.bmi09)	22.381	1	0.00	.9553783	.3071685

EPC = expected parameter change

Because we know that the modification indices are reported only if freeing a parameter would significantly improve the fit, we can see there are a lot of problems. This many problems means we need to be especially careful to avoid capitalizing on chance; we only want to free parameters that make conceptual sense. For example, if we freed the intercept at the first wave, we could reduce the chi-squared by roughly 40.95. The negative value for the expected change in the parameter estimate suggests that the linear fit slightly overestimates the initial BMI. Conceptually, the intercept is a constant value and was coded 1.0 for all years, so it would not make sense to have it vary from year to year.

There is also a big problem with the slope at bmi09 with a modification index of 83.09 and a large negative expected change. Some of the other years also have significant changes in the slope as possible improvements. As with the intercept, freeing these parameters would essentially destroy a linear trajectory that has a fixed intercept and a fixed slope.

We have several error terms we might correlate. We might expect adjacent error terms to be somewhat related and therefore correlated: e.bmi01 with e.bmi02, e.bmi02 with e.bmi03, e.bmi03 with e.bmi04, etc. In doing this, we did not blindly rely on the modification indices. We added these parameters for correlated adjacent errors because it makes sense and most of these may improve our model fit. Although the modification indices are estimated improvements one at a time, we add all these correlations in one step because it makes sense that adjacent error terms might be significantly correlated.

Before fitting a model with adjacent errors correlated, let us think about a different strategy to improve fit. We might look at our means for each year. We could run the following command to get the means for each wave with listwise deletion to remove observations with missing values from the sample:

```
. sum bmi01-bmi09 if !missing(bmi01, bmi02, bmi03, bmi05, bmi06, bmi07, ///
    bmi08, bmi09)
```

Alternatively, because we specified `method(mlmv)` to use all available information, we might want to run `sum bmi01-bmi09 if e(sample)`, where `e(sample)` uses the means for all observations for each variable, which is the information used by the estimation method. Here are those results:

```
. sum bmi01-bmi09 if e(sample)
```

Variable	Obs	Mean	Std. Dev.	Min	Max
bmi01	1365	25.41234	5.351076	14.84757	62.64201
bmi02	1287	26.099	5.792362	11.14546	61.5645
bmi03	1250	26.61132	6.053913	8.568421	62.95522
bmi05	1254	27.03011	6.102086	12.05272	67.34016
bmi06	1296	27.63835	6.681379	10.96235	76.97997
bmi07	1277	27.8068	6.501844	14.20023	62.64201
bmi08	1303	28.31192	6.662264	15.57895	69.93512
bmi09	1333	28.32258	6.481959	15.79941	64.99109

Remember, our linear growth model started at 25.63 (`Intercept`) and grew at a constant rate of 0.35 points per year. Our actual initial mean of 25.41 is a bit lower than where we started before (the modification index for the intercept is consistent with this discrepancy). The difference of means from `bmi01` to `bmi02` is 0.69 and not 0.35; from `bmi02` to `bmi03` is 0.51; from `bmi03` to `bmi05` (we do not have BMI for 2004) is 0.42 (0.21 per year); from `bmi05` to `bmi06` is 0.61; from `bmi06` to `bmi07` is 0.17; from `bmi07` to `bmi08` is 0.50; and from `bmi08` to `bmi09` is 0.01. Looking at these, we can see that the first few years, the increase in BMI is above our 0.35, but for the later years it is mostly less than this, a pattern that is consistent with a slight curvature in the growth in BMI. This is one answer to our modification indices about the intercept and slope. We might solve our problem of a weak fit by adding a quadratic term.

We could graph the means compared with a linear fit. We can enter the data directly and then create the plot as shown in figure 4.8:

```
. clear

. input year bmi

        year        bmi
1. 0 25.41234
2. 1 26.099
3. 2 26.61132
4. 4 27.03011
5. 5 27.63835
6. 6 27.8068
7. 7 28.31192
8. 8 28.32258
9. end
```

```
. twoway (connected bmi year) (lfit bmi year, lcolor(black)
> lwidth(medthick) lpattern(longdash))
```

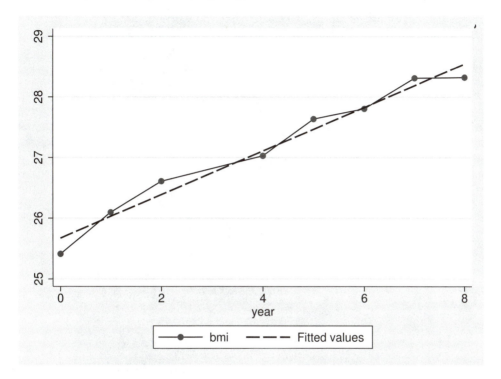

Figure 4.8. Plot of means by year

Our graph suggests a bit of leveling in the growth of BMI, and this would be tapped by adding a quadratic slope. Does adding a quadratic slope make sense? We can justify this by recognizing that BMI could not continue to go up indefinitely in a straight line. We do not know when it will start to level out, but it is possible that this could happen in the late 20s, as our means suggest.

This analysis suggests we add a latent quadratic growth factor to our model. Thus two distinct strategies are suggested to improve our model fit. One is to correlate adjacent error terms; the other is to use a quadratic. Let us fit each of these possible solutions and then combine them.

Box 4.1. Fitting a growth curve with the SEM Builder

When drawing your growth curve model with the SEM Builder, you have to do two things before you fit the model. To ensure that you will estimate the mean for the latent growth factors (the `Intercept` and the `Slope`), you need to double-click on them one at a time. When you double-click on, say, the `Intercept`, a *Variable properties* dialog box will open. In the dialog box, click on *Estimate mean*:

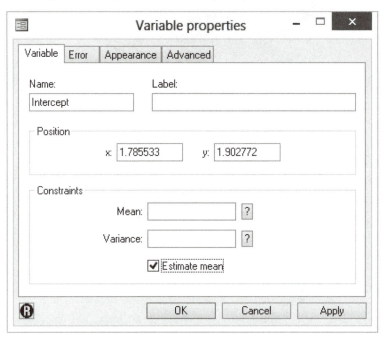

Box 4.1. (*continued*)

You next need to ensure that Stata knows all the intercepts on the observed variables are fixed at 0. Usually, the simplest way to do this is to wait until you are ready to fit the model. Before you fit it, click on **Estimation** > **Estimate**, select the **Advanced** tab, and click on the box for *Do not fit intercepts*:

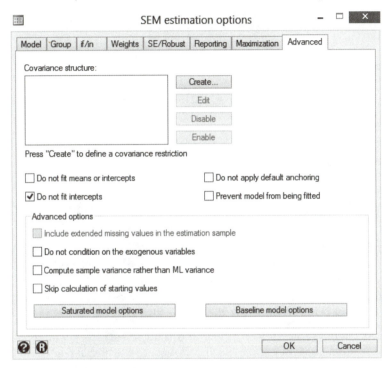

Alternatively, you could click on each observed variable and enter a 0 in the box that appears on the top toolbar.

4.4.5   Adding correlated adjacent error terms

The following is the `sem` command for correlating errors. Notice that there are no data for wave 4, so we ignore that. Stata generated names for the error terms automatically by adding an `e.` as a prefix to the observed variable name. Thus the error for `bmi01` becomes `e.bmi01`. Correlating this error term with the error term for `bmi02` is accomplished by adding the option `cov(e.bmi01*e.bmi02)`.

```
. use http://www.stata-press.com/data/dsemusr/bmiworking.dta, clear
. sem (Intercept@1 Slope@0 -> bmi01)               ///
    (Intercept@1 Slope@1 -> bmi02)                 ///
    (Intercept@1 Slope@2 -> bmi03)                 ///
    (Intercept@1 Slope@4 -> bmi05)                 ///
    (Intercept@1 Slope@5 -> bmi06)                 ///
    (Intercept@1 Slope@6 -> bmi07)                 ///
    (Intercept@1 Slope@7 -> bmi08)                 ///
    (Intercept@1 Slope@8 -> bmi09),                ///
    method(mlmv) noconstant means(Intercept Slope) ///
    cov(e.bmi01*e.bmi02) cov(e.bmi02*e.bmi03)      ///
    cov(e.bmi05*e.bmi06) cov(e.bmi06*e.bmi07)      ///
    cov(e.bmi07*e.bmi08)
. estat gof, stats(all)
. estat mindices
```

The results (not shown) for the `estat gof, stats(all)` postestimation command show a significant improvement. Our $\chi^2(26) = 196.43$, RMSEA = 0.06, and CFI = 0.99. Our previous model (without correlated error terms) had a $\chi^2(31) = 376.67$. The difference between these two chi-squared values is $376.67 - 196.43 = 180.24$. A $\chi^2(5) = 180.24$ is highly significant, $p < 0.001$. You can estimate the exact probability with `display chi2tail(`*df, chi_squared*`)`, that is, `display chi2tail(5, 180.24)`.

Rather than subtracting these by hand and using the `display` command to obtain the probability level, you can use the likelihood ratio test as implemented by the `lrtest` command. It has slightly better precision because it keeps more decimal places, and you avoid the chance of making an arithmetic mistake. After the first structural equation model that did not have correlated errors, you would tell Stata to save the results by typing `estimates store ind_errors`.[2] Then after the second structural equation model, you would save those results under a different name by typing `estimates store corr_errors`. Finally, you would run `lrtest ind_errors corr_errors`.

```
. lrtest ind_errors corr_errors
Likelihood-ratio test                              LR chi2(5) =    180.24
(Assumption: ind_errors nested in corr_errors)     Prob > chi2 =    0.0000
```

This is a significant improvement, and we could be fairly happy with this growth model of BMI during your 20s. Before we look at the results, let us consider our second alternative of adding a quadratic term.

---

2. The name you assign to each saved set of estimates is your choice.

4.4.6 Adding a quadratic latent slope growth factor

We can fit a quadratic simply by adding it as a latent variable and fixing its loading at the squared value of the corresponding linear variable. Here is our program:

```
. sem (Intercept@1 Linear@0 Quadratic@0 -> bmi01)          ///
      (Intercept@1 Linear@1 Quadratic@1 -> bmi02)          ///
      (Intercept@1 Linear@2 Quadratic@4 -> bmi03)          ///
      (Intercept@1 Linear@4 Quadratic@16 -> bmi05)         ///
      (Intercept@1 Linear@5 Quadratic@25 -> bmi06)         ///
      (Intercept@1 Linear@6 Quadratic@36 -> bmi07)         ///
      (Intercept@1 Linear@7 Quadratic@49 -> bmi08)         ///
      (Intercept@1 Linear@8 Quadratic@64 -> bmi09),        ///
      method(mlmv) noconstant means(Intercept Linear Quadratic)
. estat gof, stats(all)
. estat mindices
```

This quadratic approach also shows a substantial improvement (results not shown). The $\chi^2(27) = 177.67$, RMSEA $= 0.06$, and CFI $= 0.99$. This approach also does significantly better than the linear model with uncorrelated adjacent error terms, $\chi^2(31) = 376.67$. The difference is clearly statistically significant ($376.67 - 177.67 = 199$). A $\chi^2(5) = 199$ is highly significant, $p < 0.001$.

4.4.7 Adding a quadratic latent slope growth factor and correlating adjacent error terms

We cannot use the chi-squared values to test which of the two previous solutions is best. The model with correlated adjacent error terms was not nested in the model with a latent quadratic slope factor nor vice versa. However, both models make considerable sense and both improve the fit. What happens if we put them together?

We can do this by using the following sem command. Because this is our final model, we will look at all the results and go over them carefully.

```
. sem (Intercept@1 Linear@0 Quadratic@0 -> bmi01)
>     (Intercept@1 Linear@1 Quadratic@1 -> bmi02)
>     (Intercept@1 Linear@2 Quadratic@4 -> bmi03)
>     (Intercept@1 Linear@4 Quadratic@16 -> bmi05)
>     (Intercept@1 Linear@5 Quadratic@25 -> bmi06)
>     (Intercept@1 Linear@6 Quadratic@36 -> bmi07)
>     (Intercept@1 Linear@7 Quadratic@49 -> bmi08)
>     (Intercept@1 Linear@8 Quadratic@64 -> bmi09),
>     method(mlmv) noconstant means(Intercept Linear Quadratic)
>     cov(e.bmi01*e.bmi02) cov(e.bmi02*e.bmi03)
>     cov(e.bmi05*e.bmi06) cov(e.bmi06*e.bmi07)
>     cov(e.bmi07*e.bmi08)
```

(output omitted)

	Coef.	OIM Std. Err.	z	P>|z|	[95% Conf. Interval]	
Measurement						
bmi01 <-						
Intercept	1	(constrained)				
_cons	0	(constrained)				
bmi02 <-						
Intercept	1	(constrained)				
Linear	1	(constrained)				
Quadratic	1	(constrained)				
_cons	0	(constrained)				
bmi03 <-						
Intercept	1	(constrained)				
Linear	2	(constrained)				
Quadratic	4	(constrained)				
_cons	0	(constrained)				
bmi05 <-						
Intercept	1	(constrained)				
Linear	4	(constrained)				
Quadratic	16	(constrained)				
_cons	0	(constrained)				
bmi06 <-						
Intercept	1	(constrained)				
Linear	5	(constrained)				
Quadratic	25	(constrained)				
_cons	0	(constrained)				
bmi07 <-						
Intercept	1	(constrained)				
Linear	6	(constrained)				
Quadratic	36	(constrained)				
_cons	0	(constrained)				
bmi08 <-						
Intercept	1	(constrained)				
Linear	7	(constrained)				
Quadratic	49	(constrained)				
_cons	0	(constrained)				
bmi09 <-						
Intercept	1	(constrained)				
Linear	8	(constrained)				
Quadratic	64	(constrained)				
_cons	0	(constrained)				
mean(Interc~t)	25.48822	.1379886	184.71	0.000	25.21777	25.75867
mean(Linear)	.4935257	.0347645	14.20	0.000	.4253885	.5616628
mean(Quadra~c)	-.0177709	.0040253	-4.41	0.000	-.0256602	-.0098815

var(e.bmi01)	.4517301	.4636246			.0604317	3.376708
var(e.bmi02)	5.281952	.3642298			4.614215	6.04632
var(e.bmi03)	5.143104	.2808661			4.621054	5.724132
var(e.bmi05)	4.66403	.2825185			4.141911	5.251966
var(e.bmi06)	7.657956	.3881564			6.933752	8.457801
var(e.bmi07)	4.126451	.2330713			3.694017	4.609507
var(e.bmi08)	3.643006	.2181229			3.239625	4.096615
var(e.bmi09)	1.453963	.2716545			1.008131	2.096957
var(Intercept)	28.70621	1.158186			26.52364	31.06838
var(Linear)	.9882498	.1010058			.8088504	1.207439
var(Quadratic)	.0108736	.0011848			.0087827	.0134623
cov(e.bmi01,						
e.bmi02)	.0724345	.3078115	0.24	0.814	−.5308649	.675734
cov(e.bmi02,						
e.bmi03)	1.64931	.2262764	7.29	0.000	1.205816	2.092803
cov(e.bmi05,						
e.bmi06)	.6428412	.2578026	2.49	0.013	.1375573	1.148125
cov(e.bmi06,						
e.bmi07)	.6122682	.1976852	3.10	0.002	.2248123	.9997241
cov(e.bmi07,						
e.bmi08)	.8808256	.1806818	4.88	0.000	.5266958	1.234956
cov(Interc~t,						
Linear)	−.2281066	.2608827	−0.87	0.382	−.7394273	.2832141
cov(Interc~t,						
Quadratic)	.0200816	.0270477	0.74	0.458	−.0329309	.0730942
cov(Linear,						
Quadratic)	−.0926273	.0103805	−8.92	0.000	−.1129726	−.0722819

LR test of model vs. saturated: chi2(22)   =   61.67, Prob > chi2 = 0.0000

. estat eqgof

Equation-level goodness of fit

depvars	fitted	Variance predicted	residual	R-squared	mc	mc2
observed						
bmi01	29.15794	28.70621	.4517301	.9845075	.9922235	.9845075
bmi02	34.38599	29.10403	5.281952	.8463923	.9199958	.8463923
bmi03	35.74249	30.59938	5.143104	.8561067	.9252603	.8561067
bmi05	38.92736	34.26333	4.66403	.8801863	.9381825	.8801863
bmi06	43.43263	35.77467	7.657956	.823682	.9075693	.823682
bmi07	41.19549	37.06904	4.126451	.8998325	.948595	.8998325
bmi08	42.11323	38.47022	3.643006	.913495	.9557693	.913495
bmi09	42.01694	40.56298	1.453963	.9653958	.9825456	.9653958
overall				.9995631		

mc = correlation between depvar and its prediction
mc2 = mc^2 is the Bentler-Raykov squared multiple correlation coefficient

```
. estat gof, stats(all)
```

Fit statistic	Value	Description
Likelihood ratio		
chi2_ms(22)	61.665	model vs. saturated
p > chi2	0.000	
chi2_bs(28)	15415.481	baseline vs. saturated
p > chi2	0.000	
Population error		
RMSEA	0.034	Root mean squared error of approximation
90% CI, lower bound	0.024	
upper bound	0.044	
pclose	0.997	Probability RMSEA <= 0.05
Information criteria		
AIC	51864.285	Akaike´s information criterion
BIC	51982.333	Bayesian information criterion
Baseline comparison		
CFI	0.997	Comparative fit index
TLI	0.997	Tucker-Lewis index
Size of residuals		
CD	1.000	Coefficient of determination

Note: SRMR is not reported because of missing values.

```
. estat mindices
```

Modification indices

	MI	df	P>MI	EPC	Standard EPC
Measurement					
bmi01 <-					
Intercept	6.619	1	0.01	-.0096424	-.0095675
Quadratic	3.932	1	0.05	-3.918601	-.0756728
_cons	6.082	1	0.01	-.2390587	-.0442717
bmi02 <-					
Intercept	4.652	1	0.03	.0052842	.0048281
_cons	4.022	1	0.04	.1271397	.0216816
bmi05 <-					
Intercept	8.438	1	0.00	-.0085433	-.0073365
_cons	5.959	1	0.01	-.1867107	-.0299255
bmi07 <-					
Linear	4.747	1	0.03	-.1628344	-.0252206

cov(e.bmi01,e.bmi03)	9.680	1	0.00	2.490717	1.634075
cov(e.bmi01,e.bmi07)	6.974	1	0.01	.3794927	.2779556
cov(e.bmi01,e.bmi09)	18.079	1	0.00	-.8873363	-1.094894
cov(e.bmi02,e.bmi09)	8.224	1	0.00	.4260756	.1537491
cov(e.bmi03,e.bmi07)	7.194	1	0.01	-.4110461	-.0892256
cov(e.bmi05,e.bmi07)	8.153	1	0.00	.5748421	.1310326
cov(e.bmi08,e.bmi09)	6.805	1	0.01	.8777522	.3813865

EPC = expected parameter change

Although the model is not perfect because the chi-squared is significant, $\chi^2(22) = 61.67$, $p < 0.001$, the other measures of goodness of fit are excellent: RMSEA = 0.03 and CFI = 1.00. The intercept mean is 25.49, $z = 184.71$, $p < 0.001$; the linear slope mean is 0.49, $z = 14.20$, $p < 0.001$; and the quadratic slope mean is -0.02, $z = -4.41$, $p < 0.001$. We could write this as an equation

$$\mathrm{BMI}_{\mathrm{Est.}} = 25.49 + 0.49 \times \mathtt{year} - 0.02 \times \mathtt{year}^2$$

To simplify our interpretation, we can generate a graph of the quadratic model. We will do this by generating a small dataset that contains just nine observations for the waves of data, 0–8. We can then plot a two-way graph. We use the line graph because we are fitting the equation, and a line graph is appropriate whether we are fitting a straight line or a quadratic. The `twoway` graph command is a bit complicated because I included the labeling in the command rather than using the Graph Editor to make these changes.

```
. clear
. input year

            year
  1. 0
  2. 1
  3. 2
  4. 3
  5. 4
  6. 5
  7. 6
  8. 7
  9. 8
 10. end
. generate year_sq = year*year
. generate bmiest = 25.488 + 0.494*year -0.018*year_sq
```

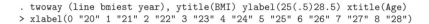

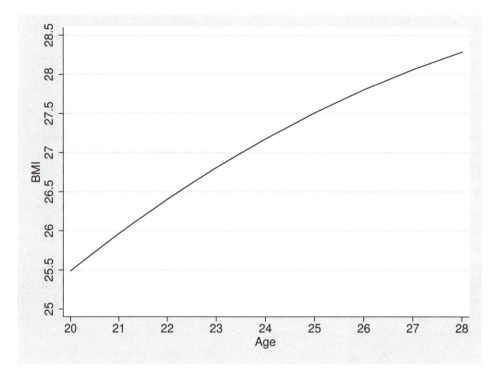

Figure 4.9. A quadratic growth curve relating BMI to age

Figure 4.9 presents the resulting quadratic graph of BMI from age 20 to age 28. You can see there is a substantial initial increase in the BMI and a slight leveling off of the effect because of the small but statistically significant quadratic term.

We need to pay special attention to the variance of the latent growth factors. The mean of the `Intercept` is 25.49, and the variance is 28.71 with a standard error of 1.16. Stata does not report a z test for the variance: the sampling distribution has a boundary problem because the variance cannot be less than 0. However, Stata reports a 95% confidence interval, and the lower limit is well above 0. Therefore, there appears to be a substantial variance in the random effects. We can take the square root of the variance to obtain a standard deviation = 5.36. Based on the assumption that the random intercepts are normally distributed, about 95% of the people will have a personal intercept within two standard deviations of the mean. Thus about 95% of the people will have an intercept between 14.77 and 36.20. This is an enormous range in the estimated initial BMI varying from underweight to morbidly obese. Whenever you have a substantial variance in the `Intercept`, you will likely want to identify exogenous variables (gender, exercise, parents' weights) that might explain this variance.

The variance around the overall linear slope of 0.49 is 0.99, so the standard deviation is 0.99. This is a huge variance relative to the size of the slope. Consider that an approximate 95% confidence interval on the slope will be $0.49 \pm 2 \times 0.99$ (95% confidence interval of -1.49 to 2.48). Some people have a steep negative slope, while others have a steep positive slope. We have a substantial variance for the random intercept and a substantial variance for the random slope. There is also variance of the random effects for the quadratic, but it is less dramatic than the variances of the random effects for the intercept and linear slope.

Box 4.2. Interpreting a linear slope when there is a quadratic slope

When you have both a linear and a quadratic slope, the interpretation of the linear slope is problematic. This is how such a growth curve might appear:

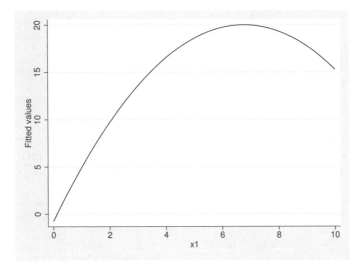

For this example, the estimated trajectory is $Y = -0.74 + 6.21X - 0.45X^2$, and both the linear slope of 6.21 and the quadratic slope of -0.45 are statistically significant. By inspecting the graph, we can see that the slope varies depending on the value of x, the wave. At wave 2, the slope is quite positive, but at wave 7 the slope is near 0, and at wave 9 the slope is quite negative.

Technically, the literal interpretation of the linear slope is the tangent of the curve at wave 0. However, the slope is a different value for any other wave. With a quadratic, you need both the linear and the quadratic slope, and interpretation is much simpler if you do a plot of the curved growth trajectory. In this example, it appears that there is a rather steep increase in y for low values of x—say, between 0 and 4 or 5. Between 5 and 7, the slope is quite flat. After that, the slope becomes negative.

Box 4.2. (continued)

You might want to be more specific. For example, you could say that initially the slope is 6.21, but it becomes flat (the highest point) at $\beta_1/(2\beta_2) = -6.21/(2 \times (-0.45)) = 6.87$. At any given wave, the slope is $\beta_1 + 2\beta_2 x$. At the last wave, the slope is $6.21 + 2 \times (-0.45) \times 10 = -2.83$. Given that the initial slope is 6.21, the slope at 6.87 is 0.0, and the slope at 10 is -2.83, you can see that the graph is easier to explain to your reader than trying to interpret the slopes directly.

4.5   How can we add time-invariant covariates to our model?

A covariate is called time-invariant if its value remains constant across waves—such as gender. In this section, we will add a pair of time-invariant covariates. We will use gender as a binary predictor of the trajectory and self-described weight during adolescence, deswgt97, as a second predictor. The self-described weight variable is coded from 1 for very underweight to 5 for very overweight. To help simplify the interpretation of the intercept, we will recode this variable, generating the variable deswgt0_97 that ranges from 0 to 4 instead of from 1 to 5. We do this so that a value of 0 is meaningful for the time-invariant covariate.

We can create a frequency distribution with Stata's tabulate command or the user-written command fre, written by Ben Jann.[3] The fre command has the advantage of including both value labels (very underweight, slightly underweight, etc.) and values $(0, 1, 2, 3, 4)$. It also includes missing values automatically and has some additional statistical information. Here is the frequency distribution of the new variable:

```
. generate deswgt0_97 = deswgt97-1
(2 missing values generated)
. label define labels 0 "Very underweight" 1 "Slightly underweight"
> 2 "About right" 3 "Slightly overweight" 4 "Very overweight"
. label values deswgt0_97 labels
. fre deswgt0_97
```

deswgt0_97

		Freq.	Percent	Valid	Cum.
Valid	0 Very underweight	39	2.34	2.34	2.34
	1 Slightly underweight	226	13.56	13.57	15.92
	2 About right	963	57.77	57.84	73.75
	3 Slightly overweight	376	22.56	22.58	96.34
	4 Very overweight	61	3.66	3.66	100.00
	Total	1665	99.88	100.00	
Missing	.	2	0.12		
Total		1667	100.00		

---

3. If you do not have fre installed, type ssc install fre to automatically download it.

When you add a time-invariant covariate, you really change the meaning of the intercept. Instead of being the expected initial BMI, it is the expected value when you are a 0 on all the covariates as well. If we include the deswgt0_97 variable as well as gender, coded so that 0 represents female and 1 represents male, then the Intercept will be the expected value of BMI at time 0 for a woman (gender = 0) who self-described her weight as very underweight (deswgt0_97 = 0).

An important alternative approach that may make it easier to interpret the intercept is to recode the quantitative variable so that a score of 0 is meaningful as the average response for our sample. We would not do this with gender because it is a nominal-level variable. If we generate a new variable defined as c_deswgt97 = deswgt97 − mean(deswgt97), then a value of c_deswgt97 = 0 would mean you perceived yourself as average when you were an adolescent. In this way, a score of 0 would be meaningful, representing the average score rather than the most extreme score on the quantitative variable. Our intercept would be the expected BMI at year 0 for a woman who perceived herself as average weight when she was an adolescent. If you do not center your quantitative variables, pay strict attention to how you interpret the resulting estimated Intercept when you have time-invariant covariates. In this case, a graphic representation would be very important.

One way to center the deswgt97 variable is as follows:

```
. egen wgtmean = mean(deswgt97)
. generate wgt_ctr = deswgt97 - wgtmean
```

A simpler alternative is to use the user-written command center, also written by Ben Jann.[4] If we run center deswgt97, this will automatically generate a new variable, c_deswgt97, that is centered. Look at what this transformation does:

```
. center deswgt97
. sum deswgt97 c_deswgt97
```

Variable	Obs	Mean	Std. Dev.	Min	Max
deswgt97	1665	3.116517	.7671892	1	5
c_deswgt97	1665	-7.36e-08	.7671892	-2.116517	1.883483

You will notice that the mean goes from 3.116517 to −0.0000000736, and the standard deviation is unchanged.

We think that men will have a higher intercept and a steeper slope than women. We also think that people who thought they were overweight as adolescents will have a larger intercept and steeper slope than those who felt they were underweight as adolescents. The steeper slope for those who thought they were overweight as adolescents represents a cumulative disadvantage that overweight people have. This is a type of interaction where the growth rate in BMI during your 20s varies depending on how overweight you reported that you were as an adolescent.

---

4. If you do not have center installed, type ssc install center to download it.

To keep this as simple as possible, we will go back to our linear growth trajectory. By using gender and perceived weight during adolescence (centered) to explain the intercept and linear slope, we are attempting to explain some of the random-effects variance.

We need to add the command (Intercept Slope <- male c_deswgt97 _cons) to make the intercept and slope dependent variables predicted by our time-invariant covariates. Here is our set of commands:

```
. sem (Intercept@1 Slope@0 -> bmi01)
>     (Intercept@1 Slope@1 -> bmi02)
>     (Intercept@1 Slope@2 -> bmi03)
>     (Intercept@1 Slope@4 -> bmi05)
>     (Intercept@1 Slope@5 -> bmi06)
>     (Intercept@1 Slope@6 -> bmi07)
>     (Intercept@1 Slope@7 -> bmi08)
>     (Intercept@1 Slope@8 -> bmi09)
>     (Intercept Slope <- male c_deswgt97 _cons)
>     if bmi01~=.|bmi02~=.|bmi03~=.|bmi05~=.|bmi06~=.|
>     bmi07~=.|bmi08~=.|bmi09~=.,
>     var(e.Intercept*e.Slope)
>     method(mlmv) noconstant
```

  (*output omitted*)

```
Structural equation model                    Number of obs      =      1581
Estimation method  = mlmv
Log likelihood     = -28750.019
```

  (*output omitted*)

		OIM				
	Coef.	Std. Err.	z	P>|z|	[95% Conf.	Interval]
Structural						
Interc~t <-						
male	1.555545	.2436739	6.38	0.000	1.077953	2.033137
c_deswgt97	3.759441	.160021	23.49	0.000	3.445806	4.073077
_cons	24.85781	.170538	145.76	0.000	24.52356	25.19206
Slope <-						
male	.0173321	.0271012	0.64	0.522	-.0357853	.0704495
c_deswgt97	.0459197	.0178505	2.57	0.010	.0109333	.0809062
_cons	.3430045	.0189564	18.09	0.000	.3058506	.3801584
Measurement						
bmi01 <-						
Intercept	1	(constrained)				
_cons	0	(constrained)				
bmi02 <-						
Intercept	1	(constrained)				
Slope	1	(constrained)				
_cons	0	(constrained)				
bmi03 <-						
Intercept	1	(constrained)				
Slope	2	(constrained)				
_cons	0	(constrained)				

```
bmi05 <-
  Intercept |         1   (constrained)
      Slope |         4   (constrained)
      _cons |         0   (constrained)

bmi06 <-
  Intercept |         1   (constrained)
      Slope |         5   (constrained)
      _cons |         0   (constrained)

bmi07 <-
  Intercept |         1   (constrained)
      Slope |         6   (constrained)
      _cons |         0   (constrained)

bmi08 <-
  Intercept |         1   (constrained)
      Slope |         7   (constrained)
      _cons |         0   (constrained)

bmi09 <-
  Intercept |         1   (constrained)
      Slope |         8   (constrained)
      _cons |         0   (constrained)
```

mean(male)	.4990512	.0125749	39.69	0.000	.474405	.5236975
mean(c_des~97)	−.0009276	.0192434	−0.05	0.962	−.0386439	.0367888
var(e.bmi01)	2.443884	.191981			2.095144	2.850673
var(e.bmi02)	4.1472	.2157233			3.745229	4.592315
var(e.bmi03)	4.772852	.2325326			4.33818	5.251077
var(e.bmi05)	5.789807	.2596509			5.302625	6.321749
var(e.bmi06)	8.228898	.3520222			7.56708	8.948599
var(e.bmi07)	3.810727	.1909167			3.454322	4.203904
var(e.bmi08)	2.922193	.1687984			2.609396	3.272487
var(e.bmi09)	3.298484	.2089123			2.913418	3.734444
var(e.Inter~t)	20.5145	.8043478			18.99706	22.15315
var(e.Slope)	.189059	.0097265			.170925	.2091168
var(male)	.2499991	.0088918			.2331651	.2680484
var(c_desw~97)	.5849257	.0208131			.5455228	.6271748
cov(e.Inter~t, e.Slope)	−.0272943	.0635517	−0.43	0.668	−.1518533	.0972647
cov(male, c_deswgt97)	−.0747626	.0098037	−7.63	0.000	−.0939775	−.0555478

```
LR test of model vs. saturated: chi2(43)   =    385.63, Prob > chi2 = 0.0000
```

```
. estat gof, stats(all)
```

Fit statistic	Value	Description
Likelihood ratio		
chi2_ms(43)	385.631	model vs. saturated
p > chi2	0.000	
chi2_bs(44)	15927.265	baseline vs. saturated
p > chi2	0.000	
Population error		
RMSEA	0.071	Root mean squared error of approximation
90% CI, lower bound	0.065	
upper bound	0.078	
pclose	0.000	Probability RMSEA <= 0.05
Information criteria		
AIC	57544.039	Akaike's information criterion
BIC	57662.086	Bayesian information criterion
Baseline comparison		
CFI	0.978	Comparative fit index
TLI	0.978	Tucker-Lewis index
Size of residuals		
CD	0.284	Coefficient of determination

Note: SRMR is not reported because of missing values.

This looks quite different than what we were doing before. We are not asking Stata to estimate the means of the Intercept or Slope when we add time-invariant covariates; we get the estimated means for the Intercept and Slope as the coefficients on _cons in (Intercept Slope <- male c_deswgt97 _cons). That is, the coefficient on _cons for the Intercept is the estimated mean for the Intercept, and the coefficient on _cons for the Slope is the estimated mean for the Slope. The sem command also allows the error term for the intercept, e.Intercept, to be correlated with the error term for the slope, e.Slope: cov(e.Intercept*e.Slope).

Finally, we are using maximum likelihood estimation with missing values. Because gender and self-described weight during adolescence have nonmissing values for a few observations that have missing values for their BMI scores for all waves, this command would have a different N than our original 1,581. To get around this, we use just those 1,581 observations that have a nonmissing value on at least one of the BMI variables: if bmi01~=.|bmi02~=.|bmi03~=.|bmi05~=.|bmi06, etc.

4.5.1 Interpreting a model with time-invariant covariates

The RMSEA = 0.07 signifies a marginal fit for our model (RMSEA between 0.06 and 0.08); the CFI = 0.98 suggests a good fit to the data. However, the $\chi^2(43) = 385.63$, $p < 0.001$, is highly significant. Remember, to simplify this model we have not included a quadratic term or correlation of adjacent errors.

As you may recall, the intercept for the linear growth curve was 25.63; for the current model, the intercept is 24.86 (this is reported as the coefficient for the _cons). The reason for the difference is that we have added the two exogenous variables to the model. The intercept is the value when all predictors are at 0. For the linear growth curve model without covariates, the intercept was simply the mean of the latent variable Intercept. However, our new model including covariates produces an equation for an individual's estimated value in the first time period based on his or her gender and self-reported weight:

$$\text{Intercept}_{\text{Est.}} = 24.86 + 1.56(\texttt{male}) + 3.76(\texttt{c\_deswgt97})$$

Thus the coefficient on _cons in the equation for the Intercept, 24.86, is the estimated intercept for a woman (male = 0) who had an average perception of her weight as an adolescent (c_deswgt97 = 0). For a comparable man, the intercept is $24.86 + 1.56 = 26.42$. This confirms our speculation that men would have a higher intercept, and the difference of 1.56 is highly significant ($z = 6.30$, $p < 0.001$).

Independent of the gender difference, the perceived weight as an adolescent also has a strong effect: 3.76, $z = 23.49$, $p < 0.001$. It may aid our understanding to think about the intercept for some scenarios. For example, consider a man who is self-described as very overweight; His response on the 0–4 scale would be a 4. From the summary command, we see the highest score for the centered variable is 1.88. Thus a man who felt he was extremely overweight as an adolescent would have an estimated Intercept (initial value) of $24.86 + 1.56 + 3.76(1.88) = 33.49$. His initial BMI value would be in the obese range of BMI scores.

You can see how gender and self-described weight explain variance in the Intercept. In the results of the estat eqgof command, we see that the $R^2 = 0.28$ for the Intercept. Thus our two time-invariant covariates are explaining 28% of the variance in the initial BMI score.

The mean Slope, when we did the simple linear growth curve with no time-invariant covariates, was 0.35. From our output, we see that the slope has an expected value of 0.34, that is, Slope ← _cons. This looks like a small difference, but we need to be very careful about interpreting slopes and intercepts when we have time-invariant covariates. Let us look at the equation for the Slope:

$$\text{Slope}_{\text{Est.}} = 0.34 + 0.02(\texttt{male}) + 0.5(\texttt{c\_deswgt97})$$

Let us consider a couple of scenarios to better understand this equation. For a woman who perceived herself to be very underweight when she was an adolescent, her original scale score on deswgt97 was 0; from the summary command, we see that this corresponds to c_deswgt97 = −2.12. Her slope equation becomes $0.34 + 0.02(0) + 0.05(−2.12) = 0.23$. By contrast, a woman who perceived herself to be very overweight when she was an adolescent would have a score on c_deswgt97 of 1.88, and her slope equation becomes $0.34 + 0.02(0) + 0.05(1.88) = 0.43$. This interaction shows us that your perception of your weight as an adolescent moderates your rate of growth; women who felt they were very underweight have a BMI increase of an estimated 0.25 points per year, and women who felt they were very overweight have a BMI increase of an estimated 0.43 points per year. In other words, women who felt they were very overweight have a BMI that increases at $0.43/0.25 = 1.72$ times the rate as women who felt they were very underweight.

We picked the lowest and highest values of c_deswgt97 in these comparisons. This was a reasonable choice because these values correspond to meaningful points, very underweight and very overweight. If we were using education that ranged from 0 to 20 years, it would probably make sense to pick less extreme values, say, one or two standard deviations below and above the mean.

The $R^2 = 0.28$ for our Intercept was substantial; the $R^2 = 0.01$ for our Slope was very small. We can conclude that gender does not show a statistically significant effect and perceived adolescent weight does show a statistically significant effect, but overall, the two time-invariant covariates do not explain a substantial amount of variance in the Slope.

4.6   Explaining the random effects—time-varying covariates

We can add time-varying covariates to our model. Unlike time-invariant covariates, time-varying covariates may have a different effect at each wave. A person's level of physical activity might vary considerably from one year to the next during his or her 20s. In this case, physical activity would be a time-varying covariate because a year in which a person was physically inactive might create a spike in BMI, while a year in which a person was extremely physically active might create a sharp dip in BMI. These changes would represent deviations from the overall growth trajectory. The effect of a time-varying covariate is illustrated in figure 4.10.

Imagine an intervention that has five waves of data starting at the end of first grade and ending at the end of fifth grade. The intervention is designed to increase positive behavior. The solid line in figure 4.10 shows the growth curve for hypothetical data, while diamonds represent the actual means at each year. At the end of first grade, the average score was 2.0. No progress was made in the second year; the average remained at 2.0. Third grade brought the students up to 2.4, fourth grade brought them to 2.8, and there was no progress in fifth grade. The intervention ended with an average score of 2.8.

It looks like a linear growth trajectory makes overall sense, but how could you explain the lack of progress in grades 2 and 5, and the substantial progress in grades 3 and 4? One possible factor would be the level of implementation of the intervention. Each school year would have had a different teacher: perhaps the teachers for grades 2 and 5 did little to implement the intervention, while the teachers at grades 3 and 4 made a huge effort to implement the intervention. We could measure the level of implementation by each teacher, and this would be a time-varying covariate. Rather than explaining the overall growth trajectory, time-varying covariates seek to explain deviations from the overall trajectory.

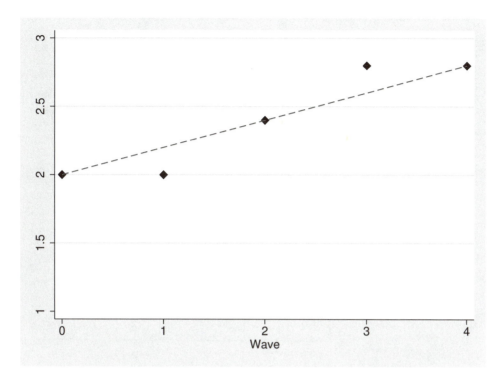

Figure 4.10. The effects of a time-varying covariate (spikes and dips)

4.6.1 Fitting a model with time-invariant and time-varying covariates

Let us explore how time-varying covariates could in be useful in our BMI example. We will simplify our figure to have just four waves of data (wave 0–wave 3). We will keep our time-invariant covariates of gender and self-perceived weight as an adolescent, and we will add a time-varying covariate: how many days the person drank in the last 30 days. Because we would expect this number to vary from year to year, it is a time-varying covariate.

The spikes and dips from the overall trajectory are exactly what time-varying covariates seek to explain. Our simplified growth curve model appears in figure 4.11. The box labeled "invariant" is a vector of time-invariant covariates (gender, race, initial health, etc.). The variables labeled `tv0`–`tv3` are the time-varying covariates. The `Intercept` and `Slope` represent the overall trajectory that may be shaped by the time-invariant covariates. By contrast, the time-varying covariates directly influence the BMI at each wave, causing a spike or a dip. In this model, time-varying covariates do not directly influence the overall trajectory, `Intercept`, or `Slope`.

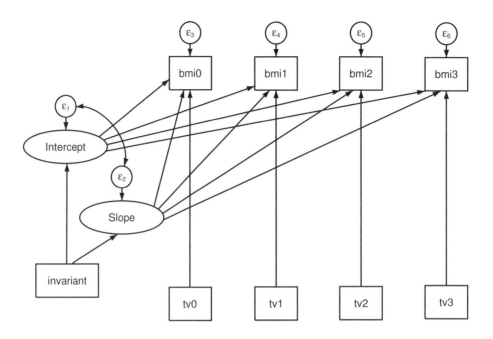

Figure 4.11. Simplified BMI model with time-invariant and time-varying covariates

Although figure 4.11 illustrates only four waves of data, we will keep all waves in our analysis. The time-varying covariate is measured at each wave, drkdays01–drkdays03 and drkdays05–drkdays09. We will first recode the extended missing value codes .a to .e as simple missing values.[5] We also center the drkdays variables before including them in the model. To fit this model, we add the time-varying covariate in the lines that specify the relationship between Intercept and Slope and BMI at each year. Here are our commands and partial results:

---

5. Stata supports several missing value codes other than the single dot. For example, .a might mean a person skipped an item and .b might mean that the person was not interviewed that year. For simplicity, we are treating all reasons as simply having missing values.

```
. recode drkdays* (.a/.e=.)
  (output omitted)

. center drkdays* deswgt97

. sem (Intercept@1 Slope@0 c_drkdays01 -> bmi01)
>     (Intercept@1 Slope@1 c_drkdays02 -> bmi02)
>     (Intercept@1 Slope@2 c_drkdays03 -> bmi03)
>     (Intercept@1 Slope@4 c_drkdays05 -> bmi05)
>     (Intercept@1 Slope@5 c_drkdays06 -> bmi06)
>     (Intercept@1 Slope@6 c_drkdays07 -> bmi07)
>     (Intercept@1 Slope@7 c_drkdays08 -> bmi08)
>     (Intercept@1 Slope@8 c_drkdays09 -> bmi09)
>     (Intercept Slope <- male c_deswgt _cons)
>     if bmi01~=.|bmi02~=.|bmi03~=.|bmi05~=.|bmi06~=.|
>     bmi07~=.|bmi08~=.|bmi09~=.,
>     cov(e.Intercept*e.Slope) method(mlmv) noconstant
  (output omitted)
```

Structural equation model   Number of obs   =   1581
Estimation method = mlmv
Log likelihood   = -53979.776

(output omitted)

	OIM Coef.	Std. Err.	z	P>|z|	[95% Conf. Interval]	
Structural						
bmi01 <-						
Intercept	1	(constrained)				
c_drkdays01	-.0227416	.0106809	-2.13	0.033	-.0436758	-.0018074
_cons	0	(constrained)				
bmi02 <-						
Intercept	1	(constrained)				
Slope	1	(constrained)				
c_drkdays02	.0071886	.0123468	0.58	0.560	-.0170107	.0313879
_cons	0	(constrained)				
bmi03 <-						
Intercept	1	(constrained)				
Slope	2	(constrained)				
c_drkdays03	-.0244292	.012214	-2.00	0.045	-.0483683	-.0004901
_cons	0	(constrained)				
bmi05 <-						
Intercept	1	(constrained)				
Slope	4	(constrained)				
c_drkdays05	-.0236272	.0124218	-1.90	0.057	-.0479735	.000719
_cons	0	(constrained)				
bmi06 <-						
Intercept	1	(constrained)				
Slope	5	(constrained)				
c_drkdays06	-.0065672	.014262	-0.46	0.645	-.0345202	.0213858
_cons	0	(constrained)				

```
bmi07 <-
    Intercept            1    (constrained)
        Slope            6    (constrained)
  c_drkdays07    -.0085683    .0125636    -0.68    0.495    -.0331925    .0160558
        _cons            0    (constrained)

bmi08 <-
    Intercept            1    (constrained)
        Slope            7    (constrained)
  c_drkdays08    -.0083243    .0109035    -0.76    0.445    -.0296947     .013046
        _cons            0    (constrained)

bmi09 <-
    Intercept            1    (constrained)
        Slope            8    (constrained)
  c_drkdays09    -.0117693    .0128466    -0.92    0.360    -.0369481    .0134095
        _cons            0    (constrained)

Interc~t <-
         male    1.606433    .2448444     6.56    0.000     1.126547    2.086319
    c_deswgt97    3.757871    .1599189    23.50    0.000     3.444436    4.071307
        _cons     24.8267    .1711425   145.06    0.000     24.49127    25.16214

Slope <-
         male    .0129514    .0274186     0.47    0.637    -.0407882    .0666909
    c_deswgt97    .0451312    .0178534     2.53    0.011     .0101391    .0801233
        _cons     .345515    .0191312    18.06    0.000     .3080185    .3830115

mean(c_drk~01)   -.2070203    .2098231    -0.99    0.324    -.6182661    .2042254
mean(c_drk~02)   -.3560257     .204503    -1.74    0.082    -.7568442    .0447928
mean(c_drk~03)   -.4028403    .2172798    -1.85    0.064    -.8287009    .0230204
mean(c_drk~05)   -.3547093    .2131742    -1.66    0.096     -.772523    .0631044
mean(c_drk~06)   -.3499527    .1963232    -1.78    0.075     -.734739    .0348337
mean(c_drk~07)   -.3455593    .1906538    -1.81    0.070    -.7192338    .0281152
mean(c_drk~08)   -.3122822     .218976    -1.43    0.154    -.7414672    .1169028
mean(c_drk~09)   -.2104118    .1987983    -1.06    0.290    -.6000494    .1792257
    mean(male)    .4990512    .0125749    39.69    0.000      .474405    .5236975
mean(c_des~97)   -.0009081    .0192431    -0.05    0.962     -.038624    .0368077
```

var(e.bmi01)	2.42794	.1917289			2.079794	2.834362
var(e.bmi02)	4.148918	.2159835			3.746478	4.594588
var(e.bmi03)	4.757472	.232338			4.323212	5.235352
var(e.bmi05)	5.761002	.2589353			5.275209	6.29153
var(e.bmi06)	8.21675	.3516084			7.555718	8.935614
var(e.bmi07)	3.815125	.191105			3.458365	4.208687
var(e.bmi08)	2.926045	.1689438			2.612969	3.276631
var(e.bmi09)	3.293869	.2088093			2.909014	3.729639
var(e.Inter~t)	20.48116	.8030823			18.96611	22.11724
var(e.Slope)	.188594	.0097073			.1704962	.2086128
var(c_drkd~01)	48.77965	2.135936			44.7679	53.1509
var(c_drkd~02)	46.2009	1.995073			42.45153	50.28141
var(c_drkd~03)	52.30925	2.278025			48.02964	56.97018
var(c_drkd~05)	48.4398	2.15292			44.39872	52.84869
var(c_drkd~06)	42.92904	1.85456			39.4438	46.72224
var(c_drkd~07)	39.16433	1.726951			35.92171	42.69966
var(c_drkd~08)	51.56922	2.267578			47.31098	56.21073
var(c_drkd~09)	42.8976	1.886437			39.35511	46.75896
var(male)	.2499991	.0088918			.2331651	.2680484
var(c_desw~97)	.5849121	.0208123			.5455107	.6271594
cov(e.Inter~t, e.Slope)	-.0256335	.0634576	-0.40	0.686	-.150008	.0987411
cov(c_drk~01, c_drkdays02)	24.42069	1.676146	14.57	0.000	21.1355	27.70587
cov(c_drk~01, c_drkdays03)	21.36572	1.726965	12.37	0.000	17.98094	24.75051
cov(c_drk~01, c_drkdays05)	16.64629	1.66109	10.02	0.000	13.39061	19.90196
cov(c_drk~01, c_drkdays06)	17.22288	1.55142	11.10	0.000	14.18215	20.2636
cov(c_drk~01, c_drkdays07)	12.61595	1.474146	8.56	0.000	9.726677	15.50522
cov(c_drk~01, c_drkdays08)	12.82834	1.706252	7.52	0.000	9.484149	16.17254
cov(c_drk~01, c_drkdays09)	13.91929	1.585489	8.78	0.000	10.81179	17.0268
cov(c_drk~01, male)	.7879629	.1066425	7.39	0.000	.5789473	.9969784
cov(c_drk~01, c_deswgt97)	-.299935	.1645307	-1.82	0.068	-.6224093	.0225394
cov(c_drk~02, c_drkdays03)	21.57399	1.689307	12.77	0.000	18.26301	24.88498
cov(c_drk~02, c_drkdays05)	17.28868	1.597517	10.82	0.000	14.1576	20.41975
cov(c_drk~02, c_drkdays06)	16.45313	1.474748	11.16	0.000	13.56268	19.34358
cov(c_drk~02, c_drkdays07)	11.54167	1.412563	8.17	0.000	8.773097	14.31024
cov(c_drk~02, c_drkdays08)	13.1205	1.688858	7.77	0.000	9.810401	16.4306
cov(c_drk~02, c_drkdays09)	13.22749	1.554854	8.51	0.000	10.18003	16.27494
cov(c_drk~02, male)	.695635	.1034319	6.73	0.000	.4929121	.8983578
cov(c_drk~02, c_deswgt97)	-.3809682	.1566147	-2.43	0.015	-.6879274	-.074009
cov(c_drk~03,						

c_drkdays05)	22.10521	1.769111	12.50	0.000	18.63782	25.57261
cov(c_drk~03, c_drkdays06)	23.48108	1.665882	14.10	0.000	20.21601	26.74614
cov(c_drk~03, c_drkdays07)	18.14174	1.586344	11.44	0.000	15.03256	21.25092
cov(c_drk~03, c_drkdays08)	17.96029	1.820459	9.87	0.000	14.39225	21.52832
cov(c_drk~03, c_drkdays09)	16.9487	1.670131	10.15	0.000	13.6753	20.2221
cov(c_drk~03, male)	.7609403	.1100569	6.91	0.000	.5452327	.9766479
cov(c_drk~03, c_deswgt97)	-.3608253	.1662481	-2.17	0.030	-.6866656	-.034985
cov(c_drk~05, c_drkdays06)	25.36452	1.65114	15.36	0.000	22.12835	28.6007
cov(c_drk~05, c_drkdays07)	21.86983	1.548059	14.13	0.000	18.83569	24.90397
cov(c_drk~05, c_drkdays08)	23.09638	1.763256	13.10	0.000	19.64046	26.5523
cov(c_drk~05, c_drkdays09)	19.55628	1.583503	12.35	0.000	16.45268	22.65989
cov(c_drk~05, male)	.5747918	.1072674	5.36	0.000	.3645516	.785032
cov(c_drk~05, c_deswgt97)	-.5583761	.165753	-3.37	0.001	-.883246	-.2335061
cov(c_drk~06, c_drkdays07)	23.6883	1.459502	16.23	0.000	20.82773	26.54887
cov(c_drk~06, c_drkdays08)	21.22191	1.653334	12.84	0.000	17.98143	24.46238
cov(c_drk~06, c_drkdays09)	20.14759	1.504842	13.39	0.000	17.19815	23.09702
cov(c_drk~06, male)	.5969232	.09906	6.03	0.000	.4027692	.7910771
cov(c_drk~06, c_deswgt97)	-.3208323	.1513803	-2.12	0.034	-.6175323	-.0241323
cov(c_drk~07, c_drkdays08)	24.48407	1.587748	15.42	0.000	21.37214	27.596
cov(c_drk~07, c_drkdays09)	21.04703	1.438187	14.63	0.000	18.22823	23.86583
cov(c_drk~07, male)	.3877053	.095625	4.05	0.000	.2002837	.575127
cov(c_drk~07, c_deswgt97)	-.3280773	.1478862	-2.22	0.027	-.617929	-.0382257
cov(c_drk~08, c_drkdays09)	24.98035	1.669741	14.96	0.000	21.70772	28.25298
cov(c_drk~08, male)	.6246036	.1103851	5.66	0.000	.4082528	.8409544
cov(c_drk~08, c_deswgt97)	-.5873023	.168524	-3.48	0.000	-.9176033	-.2570012
cov(c_drk~09, male)	.4971008	.1001582	4.96	0.000	.3007944	.6934072
cov(c_drk~09, c_deswgt97)	-.4801594	.1533831	-3.13	0.002	-.7807847	-.1795342
cov(male, c_deswgt97)	-.0747724	.0098036	-7.63	0.000	-.093987	-.0555577

LR test of model vs. saturated: chi2(99) =   456.29, Prob > chi2 = 0.0000

```
. estat eqgof
```
Equation-level goodness of fit

depvars	fitted	Variance predicted	residual	R-squared	mc	mc2
observed						
bmi01	30.9303	28.50236	2.42794	.9215029	.9599494	.9215029
bmi02	32.95927	28.81035	4.148918	.8741198	.9349438	.8741198
bmi03	34.31707	29.5596	4.757472	.8613672	.9280987	.8613672
bmi05	37.92386	32.16286	5.761002	.8480903	.9209182	.8480903
bmi06	42.1463	33.92955	8.21675	.8050422	.8972414	.8050422
bmi07	39.98002	36.1649	3.815125	.9045742	.9510911	.9045742
bmi08	41.7084	38.78235	2.926045	.9298452	.9642848	.9298452
bmi09	45.06846	41.77459	3.293869	.9269141	.9627638	.9269141
latent						
Intercept	28.48344	8.002281	20.48116	.280945	.5300425	.280945
Slope	.1897399	.0011459	.188594	.0060392	.0777126	.0060392
overall				.3031853		

mc = correlation between depvar and its prediction
mc2 = mc^2 is the Bentler-Raykov squared multiple correlation coefficient

```
. estat gof, stats(all)
```

Fit statistic	Value	Description
Likelihood ratio		
chi2_ms(99)	456.286	model vs. saturated
p > chi2	0.000	
chi2_bs(108)	16010.145	baseline vs. saturated
p > chi2	0.000	
Population error		
RMSEA	0.048	Root mean squared error of approximation
90% CI, lower bound	0.043	
upper bound	0.052	
pclose	0.789	Probability RMSEA <= 0.05
Information criteria		
AIC	108139.551	Akaike's information criterion
BIC	108622.474	Bayesian information criterion
Baseline comparison		
CFI	0.978	Comparative fit index
TLI	0.975	Tucker-Lewis index
Size of residuals		
CD	0.303	Coefficient of determination

Note: SRMR is not reported because of missing values.

4.6.2   Interpreting a model with time-invariant and time-varying covariates

The RMSEA $= 0.05$ and CFI $= 0.98$ suggest a good fit to the data. The $\chi^2(99) = 456.29$, $p < 0.001$, is highly significant. Adding the time-varying covariates greatly increases our model complexity, giving us 99 degrees of freedom compared with 43 for the model that did not include these covariates. The CFI is unchanged, while the RMSEA of 0.05 is improved and is in the very good fit range. Remember, to simplify this model, we have not included a quadratic term nor correlation of adjacent errors for BMI; we do still have correlated errors for `Intercept` and `Slope`. We also obtain estimates of the covariances for all the exogenous variables, `male`, `c_deswgt97`, and `c_drkdays01`–`c_drkdays09`.

Our intercept and slope are similar to what they were with the time-invariant covariates. We now have an `Intercept` of 24.83 compared with 24.86 before. Our slope is now 0.35 compared with 0.34 before. In the results of the `estat eqgof` command, we see that $R^2 = 0.28$ for the `Intercept`, which is the same as it was before. Our $R^2 = 0.01$, and this is also the same as before. These results should not be surprising because our model does not have a direct effect of the time-varying covariates on either the `Intercept` or the `Slope`. Such effects would be difficult to justify causally given that all but the first time-varying covariates occur after the initiation of the growth process.

How important is our time-varying covariate? Does it indicate that there are spikes in BMI related to drinking behavior in a particular year? Not so much. Only one of the effects is statistically significant, and all but one of the effects is negative. Your `bmi03` has a direct effect coming from `c_drkdays03` of -0.02, $z = -2.00$, $p < 0.05$. Thus each additional day of drinking lowers your BMI by a very small but statistically significant amount. How can we interpret this? Let us consider a person who is 23 years old and who drank on 15 of the 30 days preceding the interview. This person's BMI will be an estimated $15 \times (-0.02) = -0.3$ of a BMI point lower than what is estimated by the linear growth curve. This is how much the person who is a frequent drinker deviates from the estimated trajectory; thus the frequency of alcohol consumption does not appear to be extremely important.

Box 4.3. A note of caution when using `method(mlmv)`

We are using `method(mlmv)` in this analysis so that we can preserve all available data. When you use the default `method(ml)`, you lose a great deal of information. However, `method(mlmv)` makes assumptions that are problematic in this example: it assumes that the variables are continuous and multivariate normal. We have included gender, a binary variable, which is neither continuous nor normally distributed, although it is symmetric with a mean of 0.51. It has been argued that our method is robust against violations of normality and that the normal model applies to categorical variables measured on two levels (Graham 2009, 562). Our centered variable `c_deswgt97` is actually a categorical ordinal variable, but it is roughly normally distributed. A particular concern might be our centered `c_drkdays` variable, which has a highly skewed distribution.

If we dispense with the assumptions needed to use `method(mlmv)`, we need to have data that are missing completely at random rather than merely missing at random. We might argue that our large sample offsets the lack of normality, but Schafer and Graham (2002) indicate that a very large sample is needed for variables that are not normally distributed.

Our estimated means illustrate the potential problem with our results, even though our sample is quite large. Notice the estimated means for `c_drkdays01`– `c_drkdays09`. Each of these is centered in our observed data and has a mean of 0. Although none of the estimates is significantly less than 0, they are all in a negative direction. The estimated mean for `c_deswgt97` is approximately 0 because it is relatively more normally distributed. The mean for our observed data on `male` is 0.51, indicating that it is a symmetrical variable, and the estimated mean of 0.50 is very close.

The discrepancies in the estimated means for `c_drkdays01`–`c_drkdays09` serve as a caution because the other parameter estimates might also be affected. If you decide to switch to the default `method(ml)` for your estimator, your centering of variables would need to be computed using only the observations used with listwise deletion to obtain 0 means.

4.7 Constraining variances of error terms to be equal (optional)

For readers experienced in using the multilevel mixed-effects command `mixed` for fitting growth curves, there is one very important difference in default behavior that needs to be recognized. The `sem` command's default allows for estimated variances of the errors to be different for each time point, while the `mixed` command's default assumes that the variances of the errors are all identical. The `sem` command gives us additional information with which to evaluate our model—such as RMSEA, CFI, and the chi-squared test—that we did not have with the `mixed` command. If we followed the `mixed` default requiring equal variances, our measures of fit given by the `sem` command might be depressing.

Let us look at the estimated error variances for our measured BMI variables in the most recent example where we have included time-invariant and time-varying covariates but not equality constraints:

var(e.bmi01)	2.42794	.1917289	2.079794	2.834362
var(e.bmi02)	4.148918	.2159835	3.746478	4.594588
var(e.bmi03)	4.757472	.232338	4.323212	5.235352
var(e.bmi05)	5.761002	.2589353	5.275209	6.29153
var(e.bmi06)	8.21675	.3516084	7.555718	8.935614
var(e.bmi07)	3.815125	.191105	3.458365	4.208687
var(e.bmi08)	2.926045	.1689438	2.612969	3.276631
var(e.bmi09)	3.293869	.2088093	2.909014	3.729639

You can imagine what would happen if we forced these error variances to be equal. The variance at wave 6 is over three times greater than the variance at wave 1.

If we examine the variances of the eight error terms above, we see that there is an increasing error up to age 25 and then it drops back down. Perhaps there is something going on in a person's mid-20s that would account for this. In other situations, there might be increasing variation in errors over the course of a growth curve if the later measurements are after the completion of an intervention. On the other hand, when working with children from first grade to fifth grade, there might be a decrease in the variation of the errors as the cognitive skills of the students increase, making them better able to fully understand the questions they are asked.

Why would you force the variance of the errors to be equal? You might do this whenever you have no reason to think of them differing, because this simplifies your model considerably—you only need to estimate one variance for all the errors. Other times, this may be a practical solution to a model that does not converge. The more complicated a model gets, the harder it is for Stata to converge on a solution. When you force the error variances to be equal, you will have a worse fit to your model, but often it does not make a substantial difference.

Here is the change we make to force all the error variances to be identical:

```
. sem (Intercept@1 Slope@0 c_drkdays01 -> bmi01)
>     (Intercept@1 Slope@1 c_drkdays02 -> bmi02)
>     (Intercept@1 Slope@2 c_drkdays03 -> bmi03)
>     (Intercept@1 Slope@4 c_drkdays05 -> bmi05)
>     (Intercept@1 Slope@5 c_drkdays06 -> bmi06)
>     (Intercept@1 Slope@6 c_drkdays07 -> bmi07)
>     (Intercept@1 Slope@7 c_drkdays08 -> bmi08)
>     (Intercept@1 Slope@8 c_drkdays09 -> bmi09)
>     (Intercept Slope <- male c_deswgt97 _cons)
>     if bmi01~=.|bmi02~=.|bmi03~=.|bmi05~=.|bmi06~=.|
>     bmi07~=.|bmi08~=.|bmi09~=.,
>     cov(e.Intercept*e.Slope) method(mlmv) noconstant
>     var(e.bmi01@v e.bmi02@v e.bmi03@v e.bmi05@v
>     e.bmi06@v e.bmi07@v e.bmi08@v e.bmi09@v)
```

What is new? The last two lines of the command force all the variances to be equal by assigning a common name, v, to each of them (any common name will work). Every parameter being estimated that is assigned this common name will be forced to be equal.

Look at the error variances above, where we allowed them to be free with Stata's default. What do you think the estimates will be when we run this new **sem** command that constrains them all to have the same value? You probably will not be surprised with these results:

var(e.bmi01)	4.53166	.0748319	4.38734	4.680727
var(e.bmi02)	4.53166	.0748319	4.38734	4.680727
var(e.bmi03)	4.53166	.0748319	4.38734	4.680727
var(e.bmi05)	4.53166	.0748319	4.38734	4.680727
var(e.bmi06)	4.53166	.0748319	4.38734	4.680727
var(e.bmi07)	4.53166	.0748319	4.38734	4.680727
var(e.bmi08)	4.53166	.0748319	4.38734	4.680727
var(e.bmi09)	4.53166	.0748319	4.38734	4.680727

The parameter estimates for the `Intercept` and `Slope` do not change very much, but the goodness of fit is much worse (results not shown) with a $\chi^2(106) = 841.93$, $p < 0.001$. Still, the RMSEA $= 0.07$ and CFI $= 0.95$ are only a bit worse.

Box 4.4. Reconciling the `sem` command and the `mixed` command

Many growth models can be fit either by using the `sem` command as described in this chapter or by using the `mixed` command. One limitation of the `mixed` command is that it can only fit a single growth curve and is not able to fit parallel growth curves. The `sem` command fits growth models much faster in most cases compared with `mixed`. The `sem` command also provides additional information for assessing the fit of the model, for example, the RMSEA, CFI, and modification indices. It can be frustrating when fitting a growth model with `sem` only to find that the RMSEA is too large and the CFI is too small to meet publishable standards. If you had run the model with `mixed`, you would not have this information.

The two commands have different default treatments of error terms for each wave. The `sem` command assumes that error variances are free to differ from wave to wave; this is its default. The `mixed` command assumes that the variances of the errors for the waves are invariant as a default; this is a much more restrictive assumption. In section 4.7, we saw that the error variances for the waves varied widely from 2.43 to 8.22. When we constrained them to be invariant, they were estimated to all be 4.53. Our fit with this invariance constraint was a bit worse but still acceptable.

Is it justifiable to assume the error variances are all different? If you have a sample of adolescents being followed from age 13 to age 19 and are measuring how enmeshed they are with their families, it is possible to expect the error variances to be different; likely, they would become larger as the adolescents expand their social networks. On the other hand, it is good to think about the reasonableness of the assumption that the error variances are different from wave to wave (Grimm, Ram, and Hamagami 2011).

The `mixed` command can override the assumption of equal error variances if that is appropriate. By doing this, it will reproduce the same results as the `sem` command. Here is the `sem` command for a simple growth curve followed by the commands to reshape the data to a long format; and then the `mixed` command adds `residuals(, by(year))` to allow the error variances to be free for each wave.

```
. sem (bmi00 <- Int@1 Slope@0)              ///
      (bmi01 <- Int@1 Slope@1)              ///
      (bmi02 <- Int@1 Slope@2)              ///
      (bmi03 <- Int@1 Slope@3),             ///
      noconstant mean(Int Slope) method(mlmv)

. keep id bmi00 bmi01 bmi02 bmi03

. reshape long bmi0, i(id) j(year)

. mixed bmi0 year || id: year,               ///
    residuals(, by(year)) covariance(unstructured)
```

4.8 Exercises

1. Using the SEM Builder, draw a figure showing a linear growth curve with four waves. Draw a figure showing a quadratic growth curve with four waves. What additional parameters need to be estimated with the quadratic model?

2. In a latent growth curve, why do the arrows go from the intercept and slope to the measured scores rather than the other way around?

3. You read that between ages 75 and 85, the speed with which people perform simple math calculation slows. You are using a math speed test that has a mean of 80 and a standard deviation of 15 in the entire sample, ages 16–90. The estimated intercept for your subsample of 75- to 85-year-old people is 76 with a variance of 1.2. The estimated slope is −0.10 with a variance of 1.1.

   a. Draw a freehand figure showing the estimated growth curve for your subsample.

   b. How similar are people initially (age 75)?

   c. How similar is the rate of change for people between ages 75 and 85? Why would a fixed-effects model be problematic?

4. Given the results for exercise 3, use Stata to draw a two-way graph of the growth curve.

5. You run an **sem** linear growth curve, and your modification indices (MIs) have several estimated MIs over 3.84.

   a. What does an MI = 3.84 mean? How likely is this?

   b. The biggest MI = 117.8. You free this and refit your model. What will happen to the remaining MI estimates?

   c. Why should you generally free only one parameter at a time based on the MI values?

   d. (Optional) Give an example of a situation where you might free more than one parameter at a time.

6. Why do time-invariant covariates influence the intercept and slope but time-varying covariates influence the scores at each wave directly?

7. The **sem** approach to fitting a growth curve allows variances of errors to differ from wave to wave. Why would you constrain them to be equal? What will this constraint do to your goodness of fit?

8. Using the **exercise_wide.dta** dataset, fit a growth curve for the 12-week exercise program by using commands (**sem**, etc.).

9. Repeat exercise 8 by using the SEM Builder.

10. The `exercise_wide.dta` dataset has a time-invariant covariate, `program`. This is coded 0 for programs that increase the number of replications and is coded 1 for programs that increase the amount of weight used in each replication. Which program has better results?

5 Group comparisons

Stata offers a sophisticated way of comparing known groups with the `sem` command. In earlier chapters, we have done some types of group comparisons whenever we included a categorical variable as a predictor. In chapter 2, we fit path models; we might include a categorical variable such as gender as a predictor of one or more of the endogenous variables. This would simply add the gender effect, if any, to the estimated score on the outcome variable. For example, we might have income as an endogenous outcome variable, and education and gender as exogenous variables. We would have an effect for education, which would be added to the effect of gender to predict income. This is often referred to as an additive model. On the other hand, we might think the effects of other exogenous variables varied by gender; perhaps the effect of education on income is different for women than it is for men. A simple additive model would not be able to handle this, and we need to allow for the interaction of gender with education. We would say the effect of education on income was moderated by gender.

5.1 Interaction as a traditional approach to multiple-group comparisons

Suppose we asked a sample of likely voters their perception of a particular Republican presidential candidate for the United States. For example, just before the 2012 presidential election, we might have asked: Do you believe the U.S. is responsible to maintain the world political order? The response for each item might vary on a 7-point scale, extremely opposed to extremely supportive. Suppose we also asked a series of questions measuring the respondent's support for the Republican candidate. Finally, we could record whether the respondent was female or male.

We might predict that a) the more a study participant felt that the U.S. must be the peacekeeper for the world, the more the participant will support the Republican candidate and b) women are less likely to support the Republican candidate. An additive model appears in figure 5.1 panel a. To obtain a predicted support level for the Republican candidate, you simply add the two main effects.

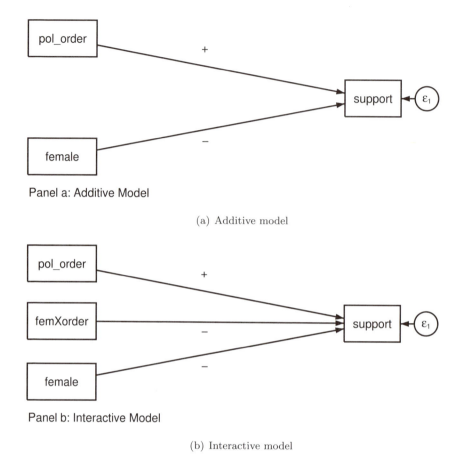

Panel a: Additive Model

(a) Additive model

Panel b: Interactive Model

(b) Interactive model

Figure 5.1. Additive and interactive models

You might believe that the additive model is too simple. Having an interventionist policy might have a positive effect on men but a negative effect on women; that is, you believe there is an interaction. Gender moderates the effect of supporting the U.S.'s responsibility for maintaining the `pol_order` of the world on `support` for the Republican candidate. The effect of one variable, `pol_order`, on another variable, `support`, is contingent on the level of the third variable, `female`. This is called an interactive model and is illustrated in figure 5.1 panel b.

We can fit a simple interactive model like this by entering a product term:

$$\text{Est. Support} = B_0 + B_1(\texttt{pol\_order}) + B_2(\texttt{female}) + B_3(\texttt{femXorder})$$

If gender results in a different effect of `pol_order` on `support`, then B_3 will be statistically significant. You probably learned how to estimate and interpret these interaction terms when you first learned multiple regression. The command is `regress support pol_order female femXorder`.[1]

5.2 The range of applications of Stata's multiple-group comparisons with sem

What does `sem` add to this traditional approach? In this chapter, we will greatly extend and enrich what we can do with a simple interaction term. For any model we have fit in this book, we can

- simultaneously fit it in two or more groups,

- test overall group differences,

- test specific group differences, and

- test for differences in means and variances as well as the size of structural paths.

5.2.1 A multiple indicators, multiple causes model

Let us take a second look at the multiple indicators, multiple causes model that was presented in chapter 3. We were interested in how important a series of factors was in producing a quality home learning environment and how the quality of the home learning environment influenced the academic achievement, educational aspirations, and teacher ratings of children. We might think that the gender of the child is an important moderator. Perhaps mother's education is more critical for daughters than it is for sons. Perhaps parental monitoring is more important for sons, but parental support is more important for daughters. Could the number of siblings have a different effect on the quality of the learning environment for sons than it does for daughters? Perhaps all these effects are equal for daughters and sons. Based on the right side of figure 3.8, we might speculate that the quality of the home learning environment has a greater or lesser effect on academic achievement, educational aspirations, and teacher ratings for girls than it does for boys.

If we were interested in the effects of divorce on child outcomes, we might compare three groups of children based on the marital status of their mothers: a) never married mothers; b) married mothers; and c) divorced mothers. The quality of the home learning environment might be more important to academic achievement for children whose parents are not married. Parental monitoring may be more important to the quality of

---

1. The interaction `femXorder` is the product of `female` and `pol_order`. When I generate an interaction as a product, I include a capital X in the middle to help me easily see that the variable is an interaction term.

the home learning environment in single parent families, whether the mother is divorced or never married, than in dual parent families.

You might think about extending the traditional approach of generating interaction terms to represent the possible moderating effects of a categorical variable on each predictor in the model, but this could quickly become too cumbersome. Instead, you can treat the categorical variable as a grouping variable and take advantage of Stata's multiple-group structural equation modeling capabilities.

5.2.2 A measurement model

Multiple-group models are also useful for comparing measurement models, which were the focus of chapter 1. Let us say that we are interested in the marital satisfaction of women and men. We might speculate that different aspects of the marriage are more central to the marital satisfaction of women than they are to the marital satisfaction of men. This is another type of interaction/moderation where the same observed indicator variable (for example, emotional support) may have a larger loading for women than it does for men, while another observed indicator variable (for example, sexual satisfaction) may have a larger loading for men than it does for women. We might also be interested in comparing the mean marital satisfaction of women and men. If we had simply taken a single satisfaction score for women and another for men and used a t test, we would not have incorporated measurement errors, and we would have implicitly assumed that all the items were equally salient to both women and men.

Perhaps we think that it may have been true in the past that women and men had very different emphases on different indicators of their perceived marital satisfaction, but not so much now. We might compare a sample from the 1980s, where this difference between women and men was stark, with a sample from the 2010s, where the difference is smaller and perhaps no longer significant. To use the date of the study (1980 versus 2010) and gender as categorical variables, we would have a four-group design: 1980 women versus 1980 men versus 2010 women versus 2010 men. Perhaps today's women place as much emphasis on sexual satisfaction as do their husbands; perhaps today's men place as much emphasis on emotional support as do their wives.

Just like any structural model we have discussed, the measurement models can be fit simultaneously for two or more groups. When comparing measurement models across groups, we are often interested in also comparing means and possibly even the variances of the latent variables.

5.2.3 A full structural equation model

We can compare full structural equation models as well as measurement models. These comparisons can evaluate both the measurement model and the structural model simultaneously. When we have a structural equation model, the first step is to compare the measurement parts of the model across our categorical variable. For example, we may want to test whether a person's belief in an activist government is an important predictor of how much he or she donates to a political party. You might measure support of an activist government by a series of items about different things the government "should" do. Should the government provide military defense, police protection, contract enforcement, abortion control, adequate housing, job opportunities, health care, or educational opportunities?

If you generate the level of support by simply adding or averaging the respondent's score on each item, you are going to miss an important difference: a strong conservative may have the same score as a strong liberal because the conservative supports the first four forms of government activism but opposes the last four, while the liberal is just the opposite. Using a variable measured this way to evaluate the role of support for an activist government is highly problematic. The same score has a fundamentally different meaning for a conservative than it has for a liberal. The structural equation modeling approach allows you to test and demonstrate invariance in the measurement model before moving on to the structural model.

Once you have evaluated the measurement model to show that the latent variables have the same or very similar meanings for each of the groups, then you can proceed to comparing the structural models. Are all of your latent predictor variables equally salient to predicting the outcome variable?

All the examples we have used so far are purely hypothetical. You might agree or disagree with any of the speculations we have proposed. The point of this introduction is to give you an idea of the rich set of research questions you can generate in your own field of study—questions that Stata's multiple-group modeling can help you resolve.

5.3 A measurement model application

Assume we are interested in the relationship between depression and a person's view of the responsibilities of government. Do people who are more depressed feel that the government should take on more responsibilities? Perhaps depressed people feel that the government should have more responsibility for social welfare than people who are not depressed.

We use data from the National Longitudinal Survey of Youth, 1997, located in the multgrp_cfa.dta dataset. We have three items about depression concerning how often a person felt blue or down, felt happy (happy should load negatively with depression), and felt depressed. We have selected 4 of the 10 items about government responsibility: providing jobs, providing the unemployed an adequate living standard, reducing income differences, and providing decent housing.

Our model appears in figure 5.2, including the results for the standardized solution using the default maximum likelihood estimator. Figure 5.2 provides the solution when both women and men are combined into one group. The figure is arranged so that we could have a large number of indicators.[2]

---

2. When you have a large number of indicators, placing the boxes for the indicators in a vertical arrangement works much better than placing them in a horizontal arrangement.

We fit this model using the SEM Builder, but the `sem` command is as follows:

```
. sem (Depress -> x1 x2 x3)                ///
      (Gov_Resp -> x4 x9 x10 x12),          ///
      standardized group(female)            /// Grouping variable is female
      ginvariant(all)                       //  All parameters constrained equal
. estat gof, stats(all)
```

The variables are `x1` (felt down or blue); `x2` (been happy); `x3` (depressed last month); `x4` (provide jobs); `x9` (help unemployed); `x10` (reduce income differences); and `x12` (provide housing).

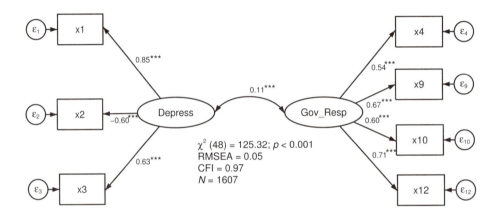

Figure 5.2. Confirmatory factor analysis for depression and support for government intervention (standardized; maximum likelihood estimator; ***$p < 0.001$)

We fit the model simultaneously in both groups and constrain all the corresponding parameters to be equal for women and men. The estimated parameters are equivalent to those from a single group model fit without the `group(female)` and `ginvariant(all)` options. We include these options here to obtain a likelihood and chi-squared statistic that can be compared with models that we fit below. Although the model has a significant chi-squared, both the root mean squared error of approximation (RMSEA) and comparative fit index (CFI) are excellent. All the loadings are highly significant and in the correct direction.[3] There is a very weak but statistically significant correlation between a person's level of depression and his or her favoring a more activist government.

_____

3. We did not reverse-code x2, so it should have a negative loading.

5.3.1 Step 1: Testing for invariance comparing women and men

Now suppose we think there might be group differences between women and men. Think about all the possible differences. The loadings on the indicator variables of the latent variables might be different for women than they are for men. Some of the indicator variables might be completely irrelevant to one group. Although not likely in this example, we might even have one or more indicators loading on a different latent variable for one of the groups.

Here we will fit a model that imposes the equivalent form on all the relationships but does not impose any equality constraints. An equivalent form solution has the same form with the same indicators loading on the latent variables for each group, but it does not require the corresponding loadings to be equal. This model places no equality constraints on the error variances, the variances of the latent variables, or the covariance of the latent variables. Additionally, we are not examining whether the means of the latent variables are the same or different. Our sem command is as follows:

```
. sem (Depress -> x1 x2 x3)
>      (Gov_Resp -> x4 x9 x10 x12),
>      group(female) ginvariant(none)  // No constraints on groups
>      means(Depress@0 Gov_Resp@0)     // Means not estimated, fixed at 0
   (output omitted)
Structural equation model               Number of obs     =      1607
Grouping variable  = female             Number of groups  =         2
Estimation method  = ml
   (output omitted)
```

	Coef.	OIM Std. Err.	z	P>|z|	[95% Conf. Interval]	
Measurement						
x1 <-						
Depress						
[*]	1	(constrained)				
_cons						
Man	3.315725	.0232397	142.67	0.000	3.270176	3.361274
Woman	3.137453	.0237937	131.86	0.000	3.090818	3.184088
x2 <-						
Depress						
Man	-.7023116	.0623569	-11.26	0.000	-.8245289	-.5800943
Woman	-.7577904	.0564539	-13.42	0.000	-.868438	-.6471428
_cons						
Man	2.169533	.024711	87.80	0.000	2.1211	2.217966
Woman	2.223203	.0235747	94.30	0.000	2.176997	2.269409
x3 <-						
Depress						
Man	.6718106	.0590329	11.38	0.000	.5561082	.787513
Woman	.7070299	.0537745	13.15	0.000	.6016339	.8124259
_cons						
Man	3.65602	.0216887	168.57	0.000	3.613511	3.698529
Woman	3.596469	.0224475	160.22	0.000	3.552473	3.640465

```
x4 <-
  Gov_Resp
      [*]              1   (constrained)
    _cons
      Man        2.412776    .0368818    65.42   0.000    2.340489    2.485063
    Woman        2.263556    .0346708    65.29   0.000    2.195603     2.33151

x9 <-
  Gov_Resp
      Man        1.091385    .0877013    12.44   0.000    .9194932    1.263276
    Woman        1.191404    .1106456    10.77   0.000    .9745427    1.408265
    _cons
      Man        2.337838    .0324588    72.02   0.000     2.27422    2.401456
    Woman        2.216898    .0328521    67.48   0.000    2.152509    2.281287

x10 <-
  Gov_Resp
      Man        1.026603    .0907875    11.31   0.000    .8486626    1.204543
    Woman        1.345166    .1251348    10.75   0.000    1.099907    1.590426
    _cons
      Man        2.309582    .0377849    61.12   0.000    2.235525    2.383639
    Woman        2.162673    .0376628    57.42   0.000    2.088856    2.236491

x12 <-
  Gov_Resp
      Man        .9152434     .076004    12.04   0.000    .7662783    1.064208
    Woman        1.056173     .097034    10.88   0.000    .8659901    1.246357
    _cons
      Man        1.793612    .0273414    65.60   0.000    1.740024      1.8472
    Woman        1.636822    .0257017    63.69   0.000    1.586448    1.687197

mean(Depress)
      [*]              0   (constrained)
mean(Gov_Resp)
      [*]              0   (constrained)

  var(e.x1)
      Man        .1179681    .0253497                     .0774201    .1797528
    Woman        .1283197    .0214042                     .0925358    .1779414
  var(e.x2)
      Man        .3384001    .0206262                     .3002949    .3813405
    Woman        .2566017    .0173433                     .2247647    .2929482
  var(e.x3)
      Man        .2377306    .0163397                     .2077687    .2720132
    Woman        .2393035    .0158903                     .2101006    .2725656
  var(e.x4)
      Man         .754767    .0453024                     .6709997    .8489917
    Woman         .710182    .0410756                     .6340708    .7954292
  var(e.x9)
      Man        .4377521     .033873                     .3761515    .5094408
    Woman        .5108531    .0348158                     .4469764    .5838582
  var(e.x10)
      Man        .7906501    .0475028                     .7028192    .8894572
    Woman        .6850604    .0460913                     .6004261    .7816246
  var(e.x12)
      Man        .3132391    .0240748                     .2694354    .3641643
    Woman        .2527114    .0216969                     .2135718    .2990239
  var(Depress)
      Man        .3216618    .0323897                     .2640509    .3918422
```

Woman var(Gov_Resp)	.3206306	.0297224			.2673614	.3845133
Man	.35249	.0475948			.2705289	.4592823
Woman	.2430548	.0378795			.1790798	.3298845
cov(Depress, Gov_Resp)						
Man	.0373598	.0156653	2.38	0.017	.0066563	.0680633
Woman	.0239425	.0132514	1.81	0.071	-.0020299	.0499148

Note: [*] identifies parameter estimates constrained to be equal across groups.
LR test of model vs. saturated: chi2(26) =   51.00, Prob > chi2 = 0.0024

. estat gof, stats(all)

Fit statistic	Value	Description
Likelihood ratio		
chi2_ms(26)	50.999	model vs. saturated
p > chi2	0.002	
chi2_bs(42)	2312.800	baseline vs. saturated
p > chi2	0.000	
Population error		
RMSEA	0.035	Root mean squared error of approximation
90% CI, lower bound	0.020	
upper bound	0.049	
Information criteria		
AIC	24812.859	Akaike's information criterion
BIC	25049.673	Bayesian information criterion
Baseline comparison		
CFI	0.989	Comparative fit index
TLI	0.982	Tucker-Lewis index
Size of residuals		
SRMR	0.030	Standardized root mean squared residual
CD	0.945	Coefficient of determination

Note: pclose is not reported because of multiple groups.

The group(female) option names the grouping variable we are using and is un-changed. What is new? The ginvariant(none) option tells Stata to apply the same form of the measurement model as defined by the first two lines but to require none of the corresponding estimates to be equal. The means(Depress@0 Gov_Resp@0) option tells Stata that the means of the latent variables are constrained to equal 0 in both groups. We do this because we are not yet ready to compare means. It is standard to work with the unstandardized solution when comparing groups, so we do not ask for a standardized solution. Also, Stata assumes for this model that women and men may have different variances on the latent variables.

These results have a $\chi^2(26) = 51$, $p < 0.001$, RMSEA $= 0.04$, and CFI $= 0.99$. The fit for this equivalent form model is convincing; all the loadings are substantial and statistically significant. If we had a poor fit to this model, we could look at the modification indices to try to figure out the difference. It might be that one pair of measurement errors is correlated for one group but not the other group. We might need to have a cross-loading in one group.

The single group model that we fit previously is nested in this model, so we can compare their fits. The model fit with `ginvariant(all)` has a $\chi^2(48) = 125.32$, while the model fit with `ginvariant(none)` (which had no constraints on any of the estimates) has a $\chi^2(26) = 51$. The difference in chi-squared is 74.32. This difference has $48 - 26 = 22$ degrees of freedom and a $p < 0.01$. An alternative way of evaluating the significance of the difference is to use a function: `display chi2tail(22, 74.32)`. This returns a probability of $1.36e-07$ or $p = 0.000000136$. Thus we can assert that a model with no invariance constraints does significantly better than a model in which all the parameters are constrained to be equal for women and men.

As for the model fit with the `ginvariant(none)` option, at the least our results allow us to assert that women and men have the same form of the model. This is a much weaker assertion than if we had required equal loadings, and without equal loadings we would weaken any additional comparisons, such as the correlation between depression and support for an activist government. Why? Because depression or support for government responsibility would have a different meaning for women than it would for men. Perhaps the government being responsible for decent housing is more central to women's beliefs about the role of government than it is to men's beliefs. In this case, it would have a significantly different loading for women and for men.

We could have fit this multiple-group model by using the SEM Builder. Open a new SEM Builder and draw your model as if you had a single group; your figure should look like figure 5.2 without any results. Using the **Select** tool, click on `Depress`. On the toolbar at the top of the window, type 0 in the **Constrain mean** box. Then select `Gov_Resp` and constrain its mean to 0 as well. Next click on **Estimation > Estimate....** On the **Group** tab, check `Group analysis`, and for the *Group variable* enter `female`. Under *Parameters that are equal across groups*, choose `None of the above`. Because we are just testing for the same form of the model, there are no constraints on parameters across groups. Now click on **OK** to get the results.

There is one big difference from results for single group models. At the center of the top bar of icons, you have a drop-down box labeled **Group** that shows 0 - Man; this indicates that the results shown are for men.

You can change this to get the results for women. Figure 5.3 shows the results for both groups. In this figure, I do not bother to indicate significance levels or reported measures of fit.

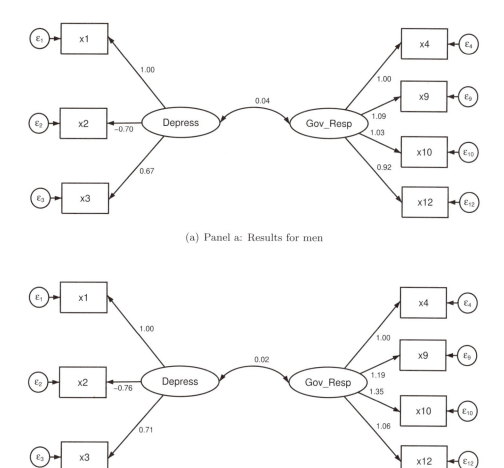

(a) Panel a: Results for men

(b) Panel b: Results for women

Figure 5.3. Same form equivalence model for women and men (unstandardized) solution

5.3.2 Step 2: Testing for invariant loadings

Based on the previous section, we will assume that some parameters are different for women than they are for men, though we do not know which parameters are different. The difference could be in the loadings, the error variances, or the covariance of the latent variables. We will begin with the loadings.

Is the restriction that the loadings are equal for women and men likely to work? If this is true, then we can assume the latent variables have the same meaning for both women and men; that is, one or another of the indicators is not more central to the meaning for women or for men. In our results above that did not impose this restriction, the unstandardized loading for x9 ← Gov_Resp is 1.09 for men versus 1.19 for women; the loading for x10 ← Gov_Resp is 1.03 for men versus 1.35 for women; etc. Some of these differences sound like they might be substantial, and some are remarkably similar. For invariant loadings to be justified, we need to impose the constraint that all the corresponding loadings are the same for women as they are for men.

We accomplish this test with the following **sem** command:

```
. sem (Depress -> x1 x2 x3)
>      (Gov_Resp -> x4 x9 x10 x12),
>      group(female) ginvariant(mcoef) // Invariant loadings
>      means(Depress@0 Gov_Resp@0)     // Means not estimated, fixed at zero
  (output omitted)
Structural equation model                  Number of obs     =       1607
Grouping variable  = female                 Number of groups  =          2
Estimation method  = ml
  (output omitted)
```

	Coef.	OIM Std. Err.	z	P>|z|	[95% Conf. Interval]	
Measurement						
x1 <-						
Depress						
[*]	1	(constrained)				
_cons						
Man	3.315725	.0231893	142.99	0.000	3.270275	3.361175
Woman	3.137453	.0238434	131.59	0.000	3.090721	3.184185
x2 <-						
Depress						
[*]	-.7334043	.0418175	-17.54	0.000	-.8153651	-.6514434
_cons						
Man	2.169533	.0248189	87.41	0.000	2.120889	2.218177
Woman	2.223203	.0234816	94.68	0.000	2.17718	2.269226
x3 <-						
Depress						
[*]	.6917124	.0397356	17.41	0.000	.6138322	.7695927
_cons						
Man	3.65602	.0217021	168.46	0.000	3.613484	3.698555
Woman	3.596469	.0224334	160.32	0.000	3.5525	3.640438
x4 <-						
Gov_Resp						
[*]	1	(constrained)				
_cons						
Man	2.412776	.0364319	66.23	0.000	2.341371	2.484182
Woman	2.263556	.0351235	64.45	0.000	2.194715	2.332397

x9 <-						
Gov_Resp						
[*]	1.138136	.0693614	16.41	0.000	1.00219	1.274082
_cons						
Man	2.337838	.0323221	72.33	0.000	2.274488	2.401188
Woman	2.216898	.0330114	67.16	0.000	2.152197	2.281599
x10 <-						
Gov_Resp						
[*]	1.169412	.0749101	15.61	0.000	1.022591	1.316233
_cons						
Man	2.309582	.0383732	60.19	0.000	2.234372	2.384792
Woman	2.162673	.0370836	58.32	0.000	2.089991	2.235356
x12 <-						
Gov_Resp						
[*]	.9808148	.0606066	16.18	0.000	.862028	1.099602
_cons						
Man	1.793612	.0273682	65.54	0.000	1.739971	1.847253
Woman	1.636822	.0256771	63.75	0.000	1.586496	1.687148
mean(Depress)						
[*]	0	(constrained)				
mean(Gov_Resp)						
[*]	0	(constrained)				
var(e.x1)						
Man	.126943	.0194284			.0940446	.1713498
Woman	.1215733	.0190382			.089442	.1652475
var(e.x2)						
Man	.334244	.0195368			.2980644	.374815
Woman	.2601488	.016571			.2296158	.2947419
var(e.x3)						
Man	.2346832	.0148303			.2073444	.2656268
Woman	.2415465	.0152703			.2133973	.273409
var(e.x4)						
Man	.7689906	.0442143			.6870367	.8607205
Woman	.6990973	.0402415			.6245118	.7825904
var(e.x9)						
Man	.4470074	.0318236			.3887903	.5139418
Woman	.5025173	.0331768			.4415233	.5719372
var(e.x10)						
Man	.7727462	.0463215			.6870882	.8690831
Woman	.7087192	.043914			.6276703	.8002337
var(e.x12)						
Man	.3101196	.022577			.2688816	.3576824
Woman	.2542512	.0199561			.2179979	.2965334
var(Depress)						
Man	.3107786	.0261338			.2635557	.3664629
Woman	.329253	.0266984			.2808715	.3859685
var(Gov_Resp)						
Man	.3114165	.0355866			.2489271	.389593
Woman	.2791954	.0314866			.2238272	.3482599

cov(Depress, Gov_Resp)						
Man	.0350302	.0145181	2.41	0.016	.0065752	.0634851
Woman	.0253985	.014217	1.79	0.074	-.0024663	.0532634

Note: [*] identifies parameter estimates constrained to be equal across groups.
LR test of model vs. saturated: chi2(31) =   56.29, Prob > chi2 = 0.0036

. estat gof, stats(all)

Fit statistic	Value	Description
Likelihood ratio		
chi2_ms(31)	56.289	model vs. saturated
p > chi2	0.004	
chi2_bs(42)	2312.800	baseline vs. saturated
p > chi2	0.000	
Population error		
RMSEA	0.032	Root mean squared error of approximation
90% CI, lower bound	0.018	
upper bound	0.045	
Information criteria		
AIC	24808.149	Akaike's information criterion
BIC	25018.052	Bayesian information criterion
Baseline comparison		
CFI	0.989	Comparative fit index
TLI	0.985	Tucker-Lewis index
Size of residuals		
SRMR	0.032	Standardized root mean squared residual
CD	0.944	Coefficient of determination

Note: pclose is not reported because of multiple groups.

What is new? We replaced `ginvariant(none)` with `ginvariant(mcoef)`. Everything else is the same. This change requires that the measurement coefficients (`mcoef`), which we have called loadings, are the same for women as they are for men.

The difference between this model and the previous model is that we have required the loadings to be invariant. The loading x2 ← `Depress` has an [*] preceding it, indicating it is constrained to be equal for both women and men. The loading is -0.73 for both groups. The loading x3 ← `Depress` is similarly constrained and is estimated as 0.69 for both groups. Although this applies to each of the loadings, a separate estimate of each error variance is shown for men and for women. Additionally, there is a separate estimate of the covariance of the latent variables `Gov_Resp` ↔ `Depress` for each group. This invariant loadings model has a $\chi^2(31) = 56.29$. This is nested in our same form model where the $\chi^2(26) = 51$, so we can compare the chi-squared values. The difference is 5.29.

We can perform a likelihood-ratio test to compare the two models. To do this, we need to fit each model, store the results of each model, and then run the likelihood-ratio test. To store results of a model we fit, we use the command `estimates store` *filename*. After the same form model with `ginvariant(none)`, we might type `estimates store form`. After the equal loadings model, we could run `estimates store loadings`. Then we would run `lrtest form loadings` to obtain our test comparing the two models. The commands for fitting both models, storing the results, and running the likelihood-ratio test follow, along with the results of the likelihood-ratio test:

```
. sem (Depress -> x1 x2 x3)           ///
      (Gov_Resp -> x4 x9 x10 x12),     ///
      group(female) ginvariant(none)  /// No constraints on groups
      means(Depress@0 Gov_Resp@0)       // Means not estimated, fixed at 0

  (output omitted)

. estimates store form                 //  Store estimates for first model
. sem (Depress -> x1 x2 x3)           ///
      (Gov_Resp -> x4 x9 x10 x12),     ///
      group(female) ginvariant(mcoef) /// Invariant loadings
      mean(Depress@0 Gov_Resp@0)        //  Means not estimated, fixed at 0

  (output omitted)

. estimates store loadings             //  Store estimates for second model
. lrtest form loadings                 //  Likelihood-ratio test
Likelihood-ratio test                            LR chi2(5)   =       5.29
(Assumption: loadings nested in form)            Prob > chi2 =      0.3815
```

The difference in degrees of freedom is 5, so the test of the invariant loadings model compared with the same form model has a $\chi^2(5) = 5.29$, $p = 0.38$. The invariant loadings model is more restrictive than the same form model and does not do significantly worse. The invariant loadings model also has excellent values for RMSEA (0.03) and CFI (0.99). These results lead us to favor the model that constrains the loadings to be equal. We can assert that there is not a statistically significant difference between women and men in the meaning of depression or government responsibility for these particular indicators.

If you had found that the equal loadings model did significantly worse than the same form equivalence model, you would reject the equal loadings model. You could proceed but would need to caution your readers that you have not justified an equal loadings model, and it appears that the latent variables have somewhat different meanings for the two groups. You can run the postestimation command `estat ginvariant, showpclass(mcoef)` to see which, if any, loadings were problematic. This command will give you a test of significance for each constrained parameter estimate. It also provides a test of significance for each parameter that was not constrained against the hypothesis that it should be constrained. Let us look at what this command reports:

```
. estat ginvariant, showpclass(mcoef)
Tests for group invariance of parameters
```

	Wald Test			Score Test		
	chi2	df	p>chi2	chi2	df	p>chi2
Measurement x1 <- Depress	.	.	.	0.389	1	0.5330
x2 <- Depress	.	.	.	0.253	1	0.6151
x3 <- Depress	.	.	.	0.006	1	0.9384
x4 <- Gov_Resp	.	.	.	2.274	1	0.1316
x9 <- Gov_Resp	.	.	.	0.352	1	0.5528
x10 <- Gov_Resp	.	.	.	3.467	1	0.0626
x12 <- Gov_Resp	.	.	.	0.019	1	0.8897

Only one of the loadings, Gov_Resp $\rightarrow$ x10, even approaches significance: $\chi^2(1) = 3.47$; $p = 0.06$. If this had been significant, then Stata would be telling us that women and men differ significantly on how important x10 is as an indicator of government responsibility.

If we can show support for the model with equal loadings for women and men, then our comparisons are much more reasonable. If each of the corresponding indicators can be constrained to have the same loading for both women and men, then both depression and government responsibility are assumed to mean the same thing for women and for men.

We could use the SEM Builder in the same way as was described at the bottom of section 5.3.1 with one difference. Instead of choosing None of the above in the *Parameters that are equal across groups* box, we would choose Measurement coefficients. The figure would show that the corresponding loadings are identical for women and men, but the covariance that was not constrained would be different. This equal loadings model does not require the error variances to be equal.

5.3.3 Step 3: Testing for an equal loadings and equal error-variances model comparing women and men

This test rarely supports both equal loadings and equal error variances. Most researchers are happy to proceed once they have demonstrated equal loadings, and some proceed even if they can only support same form equivalence. Will we be able to demonstrate equal loadings and equal error variances? Look at the separate variance estimates for the measurement errors in the results above. The variance of `e.x9` = 0.45 for women and 0.50 for men. For `e.x1`, the corresponding values are 0.13 and 0.12. Most of the variance estimates are remarkably similar, so we may be able to demonstrate equal loadings and equal error-variance equivalence. We test this model with the following:

```
. sem (Depress -> x1 x2 x3)                ///
      (Gov_Resp -> x4 x9 x10 x12),         ///
      group(female) ginvariant(mcoef merrvar)  /// Equal loadings and errors
      mean(Depress@0 Gov_Resp@0)
  (output omitted )
. estat gof, stats(all)
  (output omitted )
```

The only change is that we now require the corresponding variances of the measurement errors to be equal across groups by adding `merrvar` (constrain measurement error variances) to our `ginvariant()` option. We get a $\chi^2(38) = 74.62$, $p < 0.001$, RMSEA = 0.04, and CFI = 0.98 (results not shown). Compared with our equal loadings model, where $\chi^2(31) = 56.29$, RMSEA = 0.03, and CFI = 0.99, we seem to be doing slightly worse with the additional constraints of the equal error-variances model. The difference in chi-squared is $\chi^2(7) = 18.34$, $p < 0.05$, but the RMSEA and CFI are virtually unchanged.

If we just want a test of whether the variances of the measurement errors can be constrained to be equal across groups, then we run `estat ginvariant, class` right after running the `ginvariant(mcoef)` model. This will give us a Wald chi-squared test of whether the corresponding error variances are equal for women and men. It applies to the set of error terms. If this result is significant, then we need to recognize that there are different unique variances in the responses of women and of men to the individual items.

We could draw this model with the SEM Builder in the same way as described at the bottom of section 5.3.1 except that in the *Parameters that are equal across groups* box, we would choose `Measurement coefficients` and `Covariances of measurement errors`.

We could also constrain the variances of the latent variables, `Depress` and `Gov_Resp`, along with their covariance to be equal for women and men. The code for this simply adds `cov` to our `ginvariant()` option. This forces all parameter estimates except measurement intercepts to be equal for women and men:

```
. sem (Depress -> x1 x2 x3)            ///
      (Gov_Resp -> x4 x9 x10 x12),     ///
      group(female)                    ///
      ginvariant(mcoef merrvar cov)    /// Equal loadings, var., & covariances
      mean(Depress@0 Gov_Resp@0)
  (output omitted )
. estat gof, stats(all)
  (output omitted )
```

The model fit is remarkably good (results not shown). The $\chi^2(41) = 76.91$; $p < 0.001$, RMSEA $= 0.03$, and CFI $= 0.98$.

5.3.4 Testing for equal intercepts

So far, we have not tested for equal means on the latent variables, Depress and Gov_Resp. Before doing this, some researchers add a test of equal intercepts. When the latent means are constrained to be 0 as we have been doing so far, the intercepts, labeled _cons, are simply the means for the indicators. If the items have very different means, then this might reflect an important gender difference. For example, women might have higher means (or lower means) on the three indicators of depression. We can run tabstat to see a comparison of these means. We use the if e(sample) restriction to compute means with only those observations used to fit the models above.

```
. tabstat x1 x2 x3 x4 x9 x10 x12 if e(sample), statistics(mean) by(female)
Summary statistics: mean
  by categories of: female
```

female	x1	x2	x3	x4	x9	x10	x12
Man	3.315725	2.169533	3.65602	2.412776	2.337838	2.309582	1.793612
Woman	3.137453	2.223203	3.596469	2.263556	2.216898	2.162673	1.636822
Total	3.227754	2.196017	3.626633	2.339141	2.278158	2.237088	1.716241

We can use sem to test whether the corresponding means (intercepts) are equal for the indicators of Depress and Gov_Resp. Let us work with the model that has the invariant loadings (mcoef) but does not impose equal measurement error variances (merrvar). We can achieve this by adding the intercepts, mcons, to our ginvariant() option.

```
. sem (Depress -> x1 x2 x3)
>      (Gov_Resp -> x4 x9 x10 x12),
>      group(female) ginvariant(mcoef mcons)  // Equal loadings & coefficients
>      means(Depress@0 Gov_Resp@0)
  (output omitted )
```

		OIM				
	Coef.	Std. Err.	z	P>|z|	[95% Conf. Interval]	
Measurement						
x1 <-						
Depress						
[*]	1	(constrained)				
_cons						
[*]	3.227928	.017138	188.35	0.000	3.194339	3.261518
x2 <-						
Depress						
[*]	-.726297	.0413142	-17.58	0.000	-.8072713	-.6453226
_cons						
[*]	2.193391	.0172055	127.48	0.000	2.159669	2.227113

x3 <-						
Depress						
[*]	.6869314	.0394008	17.43	0.000	.6097072	.7641556
_cons						
[*]	3.627042	.0157728	229.96	0.000	3.596128	3.657956
x4 <-						
Gov_Resp						
[*]	1	(constrained)				
_cons						
[*]	2.334691	.0255462	91.39	0.000	2.284621	2.38476
x9 <-						
Gov_Resp						
[*]	1.131517	.0682299	16.58	0.000	.997789	1.265245
_cons						
[*]	2.274465	.0233751	97.30	0.000	2.228651	2.32028
x10 <-						
Gov_Resp						
[*]	1.166885	.0739298	15.78	0.000	1.021985	1.311784
_cons						
[*]	2.232592	.026961	82.81	0.000	2.179749	2.285434
x12 <-						
Gov_Resp						
[*]	.982076	.059958	16.38	0.000	.8645605	1.099592
_cons						
[*]	1.710201	.0190844	89.61	0.000	1.672796	1.747606
mean(Depress)						
[*]	0	(constrained)				
mean(Gov_Resp)						
[*]	0	(constrained)				

(*output omitted*)

The intercepts are now preceded by an [*] indicating their equality constraint. For example, the intercept for x1 ← Depress is 3.23. For the model that has equal intercepts and equal loadings, the $\chi^2(38) = 104.78$, $p < 0.001$. We can compare this with the model that has equal loadings but without the equal intercept constraints where we had $\chi^2(31) = 56.29$. The difference is significant: $\chi^2(7) = 48.49$, $p < 0.001$. This would lead us to reject the equal intercepts constraint. This is not especially disappointing in our example, because we want to move to comparing the means of the latent variables; if the loadings and intercepts were both equal across gender, then it would be unlikely for the latent variables to have different means.

5.3.5   Comparison of models

We can construct a table that summarizes our findings so far. Table 5.1 shows that all the models provide reasonable fits to the data.

Table 5.1. Comparison of models

Model	Chi-squared(df)	Comparison	Chi-squared(df) diff	RMSEA	CFI
1. Same form model	$51(26)$, $p < 0.01$		not applicable	0.04	0.99
2. Equal loadings model	$56.29(31)$, $p < 0.01$	2 v 1	$5.29(5)$, $p = 0.38$	0.03	0.99
3. Equal loadings and errors	$74.62(38)$, $p < 0.001$	3 v 2	$18.34(7)$, $p = 0.01$	0.04	0.98
4. Equal loadings, errors, variances, and covariances	$76.91(41)$, $p = 0.001$	4 v 5	$2.29(3)$, $p = 0.52$	0.03	0.98
5. Equal loadings and intercepts	$104.78(38)$, $p < 0.001$	5 v 2	$48.49(7)$, $p < 0.001^*$	0.03	0.99

* Tested against the equal loadings only model.

We could make a case for any of the models summarized in table 5.1 because they all provide a good fit to the data. Both the RMSEA and CFI are in the good range for all models. Usually, we would pick the most restrictive model that still provides a good fit so that when we are comparing the means for women and men, we have first eliminated as many differences as possible—other than a difference of means. I would feel very comfortable with the equal loadings model (model 2). It is not significantly worse than the completely unconstrained model (model 1). Going this way, we can assert that there is not a significant difference in how women and men rate the corresponding indicators of Depress and Gov_Resp.

You might want to argue for the equal loadings and equal error-variances model (model 3). This means that women and men weight the different indicators equally and additionally have equal unique error variances. This model provides a good overall fit in terms of the RMSEA and CFI values, but it is significantly worse than the equal loadings only model (model 2). The `estat ginvariant` command pointed to two error variances that were significantly different; we might free just those two error variances. The defense of model 3 is that its RMSEA and CFI are great and virtually as strong as for the less restricted model.

If you are willing to accept the equal loadings and equal error-variances model as reasonable, you likely would go even further and accept the model that has equal loadings, error variances, variances, and covariances (model 4). It does not do significantly worse than the less restrictive model, and its root mean squared errors and CFI are both excellent. The final model we fit, model 5, does significantly worse than model 2.

Box 5.1. Why we use unstandardized values in group comparisons

Most confirmatory factor analysis (CFA) models focus on standardized solutions like those in figure 5.2. When comparing groups, we focus on unstandardized solutions. There is a good reason for this, but it can be confusing if you are more accustomed to focusing on the standardized solution. The unstandardized parameters reflect the form of the relationship. For example, if $B = 2500$ for education predicting income, then for each year of additional education, you expect the income to be $2,500 higher. We call this the form of the relationship between education and income.

A standardized coefficient, whether it is a correlation or a β weight, measures the strength of an association. For example, the correlation between education and income might be $r = 0.60$. Are standardized or unstandardized coefficients more informative? Saying that an additional year of education increases your expected income by $2,500 has a clear and definite meaning. Saying that the correlation is 0.60 indicates a moderate to strong relationship but does not tell us what that relationship is. You can think of B as representing the form of the relationship and the r or β as reflecting the strength of the relationship. Both are useful.

Box 5.1. (*continued*)

Let us consider a simple bivariate regression of income on education. An r can be very misleading. Suppose $r = 0.70$ for women and $r = 0.50$ for men. This means the observations are closer to the estimated value for women than they are for men, but it does not mean that women have more payoff for education than do men. Suppose for men $r = 0.50$ and $B = 3000$, and for women $r = 0.70$ and $B = 1500$. Clearly, the expected payoff of additional education is much greater for men than it is for women—twice as high! The correlation for women is higher than the correlation for men because more women are closer to their estimated value than are men.

The formula linking B to β provides another way of looking at the difference in meaning. A standardized β represents a combination of B and the relative standard deviations of the two variables.

$$\beta_i = B_i \left(\frac{s_i}{s_y} \right)$$

Thus the standardized β depends on three parameters: B, s_i, and s_y. In comparing groups, we compare B's because these tell us whether the form of the relationship differs across the groups. Comparing β's will be equivalent to comparing B's only in the special case that the standard deviations are identical in each group. Thus standardized coefficients confound unstandardized comparisons by adding the relative variances of the variables.

Can you see the problem for readers who are not familiar with this distinction? If B's are invariant, then β's will only be invariant if the standard deviations for both variables are identical across groups. Thus when your test for invariance is successful but you show a standardized result, the standardized values will be different. This is not a problem: your model is invariant, and the standardized solution confounds this invariance in the form of the relationship with group differences in the standard deviations.

5.3.6   Step 4: Comparison of means

Whenever you want to compare means across groups, the multiple-group comparison model is a major advance beyond traditional t tests. It is often possible to compare means while allowing for variances of the measurement errors to be different but while imposing an equal loadings model on both groups. To estimate the means on the latent Depress and Gov_Resp variables, we fix the means of the latent variables for one group at 0, making that group the reference group. We then estimate the means for the other group(s). If the means of the other group(s) are significant, then they are significantly different from the corresponding means in the reference group.

In our example, we will illustrate this to see whether women are more or less depressed and more or less supportive of an activist government compared with men. The reference group is the group that was scored a 0 on the grouping variable; this makes men our reference group because our grouping variable, `female`, is coded 0 for men. Here is our `sem` command:

```
. sem (Depress -> x1 x2 x3)
>      (Gov_Resp -> x4 x9 x10 x12),
>      group(female) ginvariant(mcoef mcons)  // Equal loadings & intercepts
  (output omitted)
```

	Coef.	OIM Std. Err.	z	P>|z|	[95% Conf. Interval]	
Measurement						
x1 <-						
Depress						
[*]	1	(constrained)				
_cons						
[*]	3.301698	.0230346	143.34	0.000	3.256551	3.346845
x2 <-						
Depress						
[*]	-.7084887	.0413694	-17.13	0.000	-.7895713	-.627406
_cons						
[*]	2.140148	.0204886	104.46	0.000	2.099991	2.180305
x3 <-						
Depress						
[*]	.6689245	.0393142	17.01	0.000	.59187	.745979
_cons						
[*]	3.676015	.0187496	196.06	0.000	3.639266	3.712764
x4 <-						
Gov_Resp						
[*]	1	(constrained)				
_cons						
[*]	2.40612	.0303658	79.24	0.000	2.346604	2.465636
x9 <-						
Gov_Resp						
[*]	1.1288	.067956	16.61	0.000	.9956087	1.261991
_cons						
[*]	2.353155	.0293734	80.11	0.000	2.295584	2.410726
x10 <-						
Gov_Resp						
[*]	1.165547	.0737561	15.80	0.000	1.020987	1.310106
_cons						
[*]	2.315769	.0329923	70.19	0.000	2.251105	2.380432
x12 <-						
Gov_Resp						
[*]	.9842681	.0600047	16.40	0.000	.8666611	1.101875
_cons						
[*]	1.781344	.0250554	71.10	0.000	1.732237	1.830452

mean(Depress)						
Man	0	(constrained)				
Woman	-.1505065	.0326825	-4.61	0.000	-.2145629	-.08645
mean(Gov_Resp)						
Man	0	(constrained)				
Woman	-.1363504	.0323028	-4.22	0.000	-.1996628	-.073038
var(e.x1)						
Man	.1192755	.0202996			.0854446	.1665016
Woman	.1138127	.0199445			.0807284	.1604557
var(e.x2)						
Man	.3385625	.0196725			.3021197	.3794011
Woman	.2646286	.0167504			.2337532	.2995823
var(e.x3)						
Man	.238392	.0149436			.2108309	.2695562
Woman	.2454392	.0154471			.2169563	.2776613
var(e.x4)						
Man	.7688125	.0442167			.6868554	.8605488
Woman	.6990664	.0402527			.6244615	.7825843
var(e.x9)						
Man	.4497992	.0317153			.3917423	.5164602
Woman	.5050903	.0331039			.4442024	.5743243
var(e.x10)						
Man	.7728928	.046263			.6873357	.8690998
Woman	.7092482	.0438489			.6283088	.8006142
var(e.x12)						
Man	.3087283	.0226157			.2674375	.3563942
Woman	.2529446	.0199834			.2166596	.2953065
var(Depress)						
Man	.3206913	.0268366			.2721796	.3778494
Woman	.3393247	.0273009			.2898216	.3972833
var(Gov_Resp)						
Man	.3126127	.0354151			.2503665	.3903347
Woman	.2800551	.0313245			.2249238	.3486996
cov(Depress,						
Gov_Resp)						
Man	.0355964	.0146912	2.42	0.015	.0068021	.0643907
Woman	.0258739	.0143768	1.80	0.072	-.0023042	.0540519

Note: [*] identifies parameter estimates constrained to be equal across groups.
LR test of model vs. saturated: chi2(36) =   67.82, Prob > chi2 = 0.0010

. estat gof, stats(all)

(*output omitted*)

We add the option `ginvariant(mcoef mcons)`. Making the intercepts invariant by specifying `mcons` in the `ginvariant()` option and removing the `mean(Depress@0 Gov_Resp@0)` option results in Stata estimating the group means. The `mcons` option forces the intercepts to be equal, and thereby any difference in means of the indicators is reflected in the means of the latent variables. As the reference group, men have a fixed mean of 0 on each latent variable, and women have a different mean. The intercept, _cons, for each indicator has an [*] indicating that the intercept is constrained to be invariant across groups. Because we have not constrained the variances of the measurement errors to be invariant, our model is equal loadings rather than equal loadings and equal error variances.

Our model has $\chi^2(36) = 67.82$, $p < 0.01$, RMSEA $= 0.03$, and CFI $= 0.99$. The asterisks show where we get a single estimate for both women and men. All the unstandardized loadings are constrained across groups, as are the intercepts. The variances and covariances are not constrained for either the measurement model or the latent variables, so we get separate estimates for women and men. By default, the mean level of depression for men is fixed at 0, and the mean on `Depress` for women is -0.15, $z = -4.61$, $p < 0.001$. Women are significantly less depressed, on average, than men.

Is this a big difference? It is difficult to provide a simple interpretation of the mean of the latent variable because its value depends on the scale on which it is being measured. Many journals expect us to report a measure of effect size. Although I am unaware of a standard approach for estimating the effect size, here is one possibility that seems reasonable. The variances for men and women on `Depress` are 0.32 and 0.34, so the standard deviations are 0.57 and 0.58, respectively. We can average these two standard deviations (`display (0.57 + 0.58)/2`) to get a pooled estimate of the standard deviation of 0.58. (If we had substantially different sample sizes for the two groups, we might want to use a weighted average.) As a measure of effect size, we can divide our mean of -0.15 by our pooled standard deviation of 0.58:

$$\text{Effect Size} = \frac{\text{Latent Mean}}{\text{Pooled Standard Deviation}}$$
$$= \frac{-0.15}{0.58}$$
$$= -0.26$$

This would be a small to moderate difference in `Depress` for women and men, with women scoring significantly lower on depression than men. The average woman's score on depression is just over a quarter of a standard deviation below the average man's score. The same formula can be used to estimate the effect size for the support for an activist government latent variable.

To fit this model using the SEM Builder, we would add a constraint on the measurement intercepts and remove the constraint on the latent means. Using the **Select** tool, we click on `Depress`. In the **Constrain mean** box on the toolbar, we remove the 0 constraint. Then click on `Gov_Resp`, and remove the 0 constraint for it as well. Now click on **Estimation > Estimate...**. On the **Group** tab, choose `Measurement coefficients` and `Measurement intercepts` in the box labeled *Parameters that are equal across groups*. Click on **OK** to fit the model.

We had our SEM Builder set up to only show measurement parameters (loadings) and the covariance of the latent variable (see appendix A for more details on how to do this). We now would like to see the mean and variance of the latent variables for the two groups. To do this, click on **Settings > Variables > Latent Standard Exogenous...**. On the **Results** tab (see figure 5.4), make the first and second boxes under *Exogenous variables* read `Custom`. To the right of the first box, enter `M = {\auto}{\mean}`. To the right of the second box, enter `V = {\auto}{\var}`. This is more involved than using the default settings but will label the values as `M` for mean and `V` for variance. Figure 5.5 shows the resulting model.

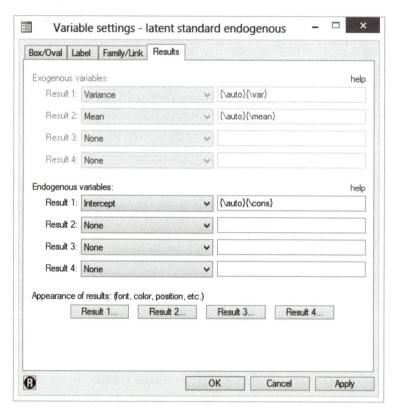

Figure 5.4. Variable settings dialog box

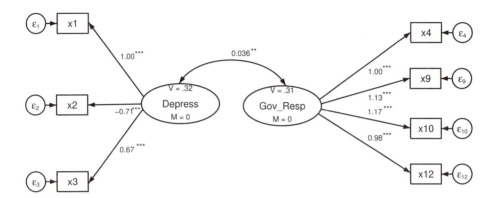

(a) Panel A: Results for men (means fixed at 0, reference means)

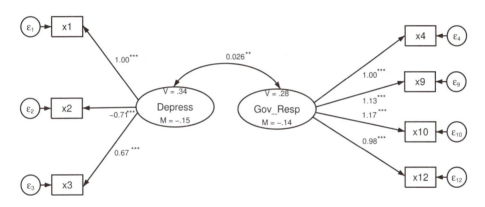

(b) Panel B: Results for women

Figure 5.5.   Comparison of means on depression and government responsibility for women and men with an equal loadings measurement model ($**p < 0.01$; $***p < 0.001$)

We could apply the same tests of mean differences assuming equal loadings and error variances on a same form measurement model. For only assuming same form equivalence, our interpretation of the difference of means is less straightforward because `Depress` and `Gov_Resp` would have different meanings for women than for men, making a direct comparison somewhat problematic. Traditional tests for mean differences ignore this possible problem. Here are the commands (notice that we need the invariant intercepts, `mcons`, for each of these models):

Same form equivalence:

```
. sem (Depress -> x1 x2 x3)          ///
    (Gov_Resp -> x4 x9 x10 x12),     /// Mean comparison with
    group(female) ginvariant(mcons) //  Same form equivalence
. estat gof, stats(all)
```

Equal loadings and error variances:

```
. sem (Depress -> x1 x2 x3)               /// Mean comparison with
    (Gov_Resp -> x4 x9 x10 x12),          /// Equal loadings and
    group(female)                         /// Equal errors
    ginvariant(mcoef merrvar mcons)       //
. estat gof, stats(all)
```

The results are not presented here, but table 5.2 summarizes the measures of fit produced by `estat gof` for the three levels of equivalence. The model for comparison of means that assumes equal loadings and error variances does significantly worse than the equal loadings model in terms of chi-squared but not in terms of RMSEA or CFI, both of which are excellent for both models. You would be safe going with either the equal loadings model or the equal loadings and error-variances model based on these results.

Table 5.2. Comparison of models

Model testing mean differences	Means for women	Chi-squared(df)	Chi-squared(df) diff	RMSEA	CFI
Same form equivalence model		$63.02(31)$, $p = 0.001$	not applicable	0.04	0.99
Depress	-0.15^{***}				
Gov_Resp	-0.13^{***}				
Equal loadings model		$67.82(36)$, $p = 0.001$	$4.80(5)$, $p = 0.44$	0.03	0.99
Depress	-0.15^{***}				
Gov_Resp	-0.14^{***}				
Equal loadings and errors model		$86.02(42)$, $p < 0.001$	$18.20(7)$, $p = 0.01$	0.04	0.98
Depress	-0.15^{***}				
Gov_Resp	-0.14^{***}				

$^{***}p < 0.001$

5.3.7   Step 5: Comparison of variances and covariance of latent variables

When we compare groups, we are usually interested in comparing means, but sometimes we may be interested in comparing variances and, in the case of two or more latent variables, comparing covariances. In this example, we are interested in the covariance of `Depress` with `Gov_Resp`. We will estimate this without comparing mean differences. Is the covariance stronger for women than it is for men? Let us illustrate this with constraints that the loadings are equal (`mcoef`) and that the error variances of the indicators are also equal (`merrvar`).

We fit two models: the first allows the covariance to be different for women than it is for men, and the second constrains the covariance to be equal. If the model constraining the covariance to be equal has a significantly worse fit, then we can assert that the co-variances are significantly different. On the other hand, if the model that constrains the covariance to be equal is not a significantly worse fit, then we assert that the covariances are not significantly different. We test this by using the unstandardized model with a covariance but then replay the standardized results that convert the covariance into a correlation.

Here are our commands:

Equal loadings, variances and covariance unconstrained:

```
. sem (Depress -> x1 x2 x3)                  ///
      (Gov_Resp -> x4 x9 x10 x12),           ///
      group(female) ginvariant(mcoef merrvar)  /// Equal loadings/errors
      mean(Depress@0 Gov_Resp@0)             //  Equal means
.  sem, standardized                         //  Standardized reports r
```

Equal loadings, variances unconstrained, covariance constrained to be equal:

```
. sem (Gov_Resp -> x4 x9 x10 x12)            ///
      (Depress -> x1 x2 x3),                 ///
      group(female) ginvariant(mcoef merrvar) /// Equal loadings/errors
      mean(Depress@0 Gov_Resp@0)             /// Equal means
      cov(Depress*Gov_Resp@a)                //  Equal covariance
. sem, standardized                          //  Standardized reports r
```

Table 5.3 summarizes our results. When we allow the covariance to be different, we obtain a small but significant correlation between `Depress` and `Gov_Resp` for men ($r = 0.11$, $p < 0.05$) and a slightly smaller and only marginally significant correlation for women ($r = 0.09$, $p = 0.06$). The model provides a good fit to the data: $\chi^2(38) = 74.62$, $p < 0.001$, RMSEA $= 0.04$, CFI $= 0.98$. The fact that one of the covariances is significant and the other is only marginally significant ($p < 0.10$) does not mean they are significantly different from each other. When we constrain our results to have the same covariance for women and men, the resulting covariance $= 0.03$, $p < 0.01$. Thus there is a weak but statistically significant relationship with more depressed people tending to be more supportive of increased governmental responsibility. There is insufficient evidence to demonstrate that the covariance differs significantly for women and men.

Table 5.3. Model testing of covariance difference

Model testing covariance Difference (equal loadings)	Covariance	Correlation	Chi-squared (df)	Chi-squared(df) difference	RMSEA	CFI
Covariance unconstrained			74.62(38), $p < 0.001$		0.04	0.98
Men	0.04*	0.11*				
Women	0.03	0.09				
Covariance constrained			74.82(39), $p < 0.001$	$0.20(1)$, $p = $ ns	0.03	0.98
Men	0.03**	0.10**				
Women	0.03**	0.10**				

**$p < 0.01$; ***$p < 0.001$

When we round to two decimal places, the correlations are equal in the model where the covariances are constrained to be equal, but had we kept additional decimal places, the correlations would not have been equal. Box 5.1 explains why standardized results such as correlations might not be equal even though the corresponding unstandardized values are equal. It is important to test for the constraint using the unstandardized solution and the difference in chi-squared.

These results can be extended to compare variances of the latent variables in the same fashion as we compare covariances. You fit the model twice: once with the variances having a label—for example, @a or @b—and the second time without these labels so that the estimates can vary for the two groups. You then use a chi-squared difference test to test whether the difference is significant. We have no hypothesis about a variance difference in this example. However, if we wanted to test whether the variance in the Depress latent variable was greater for one group or the other, we would enter var(Depress@c Gov_Resp). If we wanted to test for both variances being simultaneously equal, we would enter var(Depress@c Gov_Resp@d).

5.4 Multiple-group path analysis

Multiple-group analysis of path models focuses on the structural coefficients, the unstandardized paths in your model. Which paths are significantly different between groups and which paths can be treated as equal? We can apply this to a path model from virtually any field of study because there usually is a category of people who have different structural relationships. That is, some predictors will be more or less important for one group than they are for the other group(s). Women may differ from men on financial payoff for additional education. Obese people may differ from people who are not obese. Republicans may differ from Democrats, and both may differ from independents. Older people may differ from younger people, and both may differ from middle-age people.

We will explore a simple path model that compares Head Start children who live in a foster home with Head Start children who do not live in a foster home. Colleagues of mine were interested in predicting problem behavior among first graders based on three variables using a national sample (U.S. Department of Health and Human Services 2010). The quality of the preschool teacher–child relationship, tcr, and the quality of the preschool based on systematic observation, qual, are used as predictors of behavioral problems the child had in kindergarten, bk. The first pair of predictors, tcr and qual, were felt to lead to the level of kindergarten problem behavior. All three variables were thought to lead to the level of first grade problem behavior, b1.

The full study was much more complex than this, and we are using only a small subset of the data. The results we report will differ from those in the larger and more complex study based on the entire sample. We are just using this small subsample of data to illustrate how to compare path models across groups. Do not draw any substantive conclusions from the results we report here. Our subset of the data is in the multgrp_path.dta dataset. We have a grouping variable, grp, that is coded 0 for Head

Start children who do not live with a foster parent and 1 for Head Start children who live with a foster parent. Figure 5.6 shows our model, where `qual` is quality of preschool, `tcr` is preschool teacher–child relationship, `bk` is behavioral problems in kindergarten, and `b1` is behavioral problems in first grade.

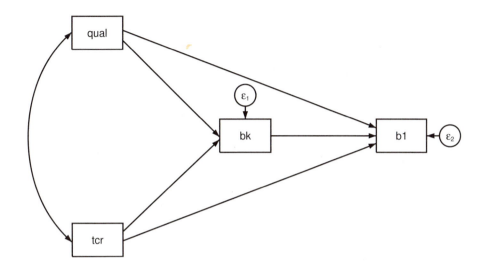

Figure 5.6. Path model

We can fit this model with all the children in our subsample and ignore whether they come from a foster home situation, but we are fundamentally interested in possible group differences. If we simply identify our grouping variable, `grp`, we obtain solutions that allow all the parameters being estimated in our path model to be different for our two groups. This is the same idea for the structural model as the same form equivalence was for the measurement model. That is, we allow the path coefficients (structural parts) and the error variances to differ across groups. Here is our command:

```
. sem (bk <- qual tcr) (b1 <- qual tcr bk),
>     group(grp) cov(tcr*qual)
```
  (*output omitted*)

```
Structural equation model                    Number of obs    =       175
Grouping variable  = grp                      Number of groups =         2
Estimation method  = ml
Log likelihood     = -1673.6574
```

		OIM				
	Coef.	Std. Err.	z	P>|z|	[95% Conf.	Interval]
Structural						
bk <-						
qual						
Not Foster	-.1907771	.2448421	-0.78	0.436	-.6706588	.2891046
Foster	.5022195	.2083046	2.41	0.016	.0939499	.9104891
tcr						
Not Foster	-.098208	.0311103	-3.16	0.002	-.159183	-.037233
Foster	-.1479874	.023454	-6.31	0.000	-.1939564	-.1020184
_cons						
Not Foster	9.975651	3.008554	3.32	0.001	4.078994	15.87231
Foster	7.862935	2.321357	3.39	0.001	3.31316	12.41271
b1 <-						
bk						
Not Foster	.2972905	.0766622	3.88	0.000	.1470352	.4475457
Foster	.267999	.0952347	2.81	0.005	.0813425	.4546555
qual						
Not Foster	-.0931447	.1726518	-0.54	0.590	-.431536	.2452466
Foster	-.0972529	.1951913	-0.50	0.618	-.4798209	.285315
tcr						
Not Foster	-.0548214	.023119	-2.37	0.018	-.1001337	-.009509
Foster	-.0226844	.0255468	-0.89	0.375	-.0727551	.0273864
_cons						
Not Foster	5.414792	2.247957	2.41	0.016	1.008877	9.820707
Foster	3.552939	2.237908	1.59	0.112	-.8332796	7.939157

  (*output omitted*)

var(e.bk)						
Not Foster	6.214582	.9589308			4.592706	8.409209
Foster	5.205083	.7716529			3.892577	6.960144
var(e.b1)						
Not Foster	3.067992	.4734014			2.26731	4.151427
Foster	4.29595	.636874			3.212689	5.744466

  (*output omitted*)

```
LR test of model vs. saturated: chi2(0)   =        0.00, Prob > chi2 =      .
```

The `group()` option identifies the grouping variable, `grp`, which has a value of 0 for children who are not from foster homes and a value of 1 for children who are from foster homes. We have asked for an unstandardized solution (the default) because these are best for comparing groups (see box 5.1). Adding the covariance option, `cov(tcr*qual)`, will not change the results of the other parameter estimates but will force Stata to include the covariance estimate in the results (we may want to have this in the figure).

You can see that there is a separate parameter estimate for each path depending on the group the child was in. Both groups were estimated simultaneously with no equality constraints across the groups. For example, the path coefficient for `qual → bk` (quality of the child's preschool to behavioral problems in kindergarten) is insignificant for children who are not from foster homes ($B = -0.19$, $z = -0.78$, $p = 0.44$); however, this same path coefficient is significant and positive for children from foster homes ($B = 0.50$, $z = 2.41$, $p = 0.02$). It appears that the quality of a preschool is a significant predictor of behavioral problems in kindergarten, but only if the child is from a foster home situation. Several of the predictors have quite different path coefficients for the two groups. At the bottom of the table, we see our $\chi^2(0) = 0.00$, meaning that our model is just-identified.

We also run one postestimation command, `estat ggof`. This additional command gives us the sample size for each group and the chi-squared test for goodness of fit for each group.

```
. estat ggof
Group-level fit statistics
```

	N	SRMR	CD	chi2	df	p>chi2
grp						
Not Foster	84	0.000	0.165	0.000	0	.
Foster	91	0.000	0.347	0.000	0	.

We see that we have 84 children who are not from foster homes and 91 children who are. The overall model as well as the solutions for each group have a chi-squared of 0 because our model is just-identified within each group—that is, we have no degrees of freedom.

We could have fit this path model by using the SEM Builder. After drawing the model, click on **Estimation > Estimate...**, and then click on `Group analysis`. Enter `grp` as the *Group variable*. If you want a standardized solution, you can move to the **Reporting** tab and click on *Display standardized coefficients and values*. Click on **OK**. Your results are shown for the group scored 0, that is, not a foster child. You can change to the foster child group by selecting it at the middle of the top bar. Figure 5.7 shows how you might present the standardized results after deleting some of the coefficients you do not want to show.

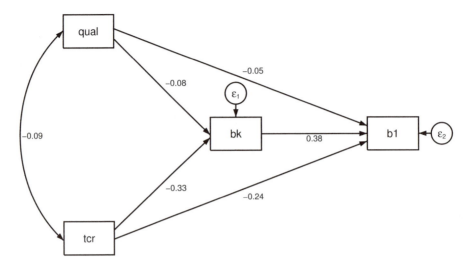

(a) Panel A: Not a foster child

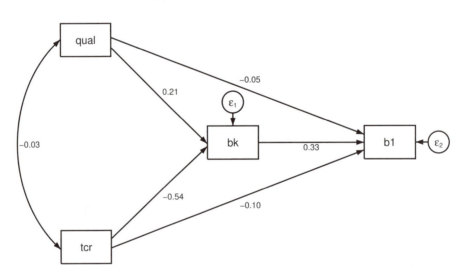

(b) Panel B: Foster child

Figure 5.7. Standardized results for children who are not foster children ($N = 84$) and who are foster children ($N = 91$)

5.4.1   What parameters are different?

The parameters appear to be different between foster children and nonfoster children. Which of these differences is statistically significant? We can run the postestimation command `estat ginvariant`, which will tell us which parameters are significantly different.

```
. estat ginvariant
Tests for group invariance of parameters
```

	Wald Test			Score Test		
	chi2	df	p>chi2	chi2	df	p>chi2
Structural						
bk <-						
qual	4.647	1	0.0311	.	.	.
tcr	1.632	1	0.2014	.	.	.
_cons	0.309	1	0.5782	.	.	.
b1 <-						
bk	0.057	1	0.8106	.	.	.
qual	0.000	1	0.9874	.	.	.
tcr	0.870	1	0.3510	.	.	.
_cons	0.345	1	0.5572	.	.	.
(output omitted)						
var(e.bk)	0.673	1	0.4121	.	.	.
var(e.b1)	2.395	1	0.1218	.	.	.

(output omitted)

These results indicate that the invariance constraint on `qual` $\rightarrow$ `bk` is significant: $\chi^2(1) = 4.65$; $p = 0.03$. This suggests that the difference is real for this path. We should temper our interpretation of these chi-squared tests because they have no adjustment for multiple comparisons. The quality of the preschool as a predictor of kindergarten problems is significantly different for children from foster families. Thus it appears that the quality of the preschool has a surprising significant effect on behavioral problems, whereas for nonfoster children, the effect is negative and not statistically significant. None of the paths going to first grade behavioral problems are significantly different. School quality matters more for children from foster homes than it does for other children from Head Start programs.

Box 5.2. Why does it matter if groups have different N's?

There is a parallel to the problem of unequal Ns when doing a two-sample t test for a difference of means. Suppose the unstandardized effect for a path is 0.10 in a large group, $N = 10000$, but the same path is 2.10 for a small group, $N = 50$. When you fit the multiple-group model with the constraint that the path is the same for both groups, you are pooling your data. There are 10,050 people in your sample, and whatever fits 10,000 of them will drown out the effect for the other 50. Perhaps an unstandardized path of 0.15 is the constrained value for both groups. The pooled estimate will surely be weighted so that it is much closer to the sample of 10,000 than it is to the sample of 50. With unequal N's, we may get misleading constrained values, and our tests of differences may be problematic.

If somebody were to replicate your study with 5,000 in one group and 5,000 in the other group, the constrained unstandardized path would be approximately in the middle, say, around 1.1. The relative sizes of the groups has a huge effect on the constrained estimate.

A second problem is that results may be hard to explain to your readers. The large group of 10,000 with the path of 0.10 will likely be statistically significant. This is because a large sample has the power to demonstrate statistical significance even when the effect is subtle. By contrast, the small group of 50 people with a path of 2.10 might not be statistically significant because the small sample gives you very little power to demonstrate statistical significance.

We need to exercise great caution when we do a multiple-group analysis with substantially unequal sample sizes.

We can refit the model and force all the structural parameter estimates to be equal except for the effect of preschool quality on behavioral problems in kindergarten. The command is a bit more complicated. We add a prefix to identify the group, a `0:` for nonfoster children and a `1:` for foster children. We then assign the same label to all the parameter estimates we are constraining to be equal across groups. Thus we assign `@a1` for `bk ← tcr` for both group 0 and group 1. We do not attach a label for school quality because we are going to allow that to differ across groups. As before, we add `cov(tcr*qual)` to obtain an estimate of the covariance of the exogenous variables. We have not put any equality constraints on the variances or the covariance.

Here is our command:

```
. sem (0: bk <- qual tcr@a1)
>     (1: bk <- qual tcr@a1)
>     (0: b1 <- qual@b1 tcr@c1 bk@d1)
>     (1: b1 <- qual@b1 tcr@c1 bk@d1),
>     group(grp) cov(tcr*qual)
```
  (*output omitted*)

```
Structural equation model                    Number of obs      =        175
Grouping variable   = grp                     Number of groups   =          2
Estimation method   = ml
Log likelihood      = -1675.1843
 ( 1)  [b1]0bn.grp#c.bk - [b1]1.grp#c.bk = 0
 ( 2)  [bk]0bn.grp#c.tcr - [bk]1.grp#c.tcr = 0
 ( 3)  [b1]0bn.grp#c.qual - [b1]1.grp#c.qual = 0
 ( 4)  [b1]0bn.grp#c.tcr - [b1]1.grp#c.tcr = 0
```

		OIM				
	Coef.	Std. Err.	z	P>|z|	[95% Conf. Interval]	
Structural						
bk <-						
qual						
Not Foster	-.2120664	.2457966	-0.86	0.388	-.6938188	.2696861
Foster	.5074561	.2089362	2.43	0.015	.0979487	.9169635
tcr						
[*]	-.1300159	.0189712	-6.85	0.000	-.1671988	-.0928331
_cons						
Not Foster	12.23906	2.454529	4.99	0.000	7.428268	17.04985
Foster	6.673325	2.138828	3.12	0.002	2.481299	10.86535
b1 <-						
bk						
[*]	.2781014	.0588937	4.72	0.000	.1626718	.3935309
qual						
[*]	-.0948936	.1279495	-0.74	0.458	-.34567	.1558828
tcr						
[*]	-.0385108	.0170227	-2.26	0.024	-.0718747	-.0051468
_cons						
Not Foster	4.396446	1.608635	2.73	0.006	1.243578	7.549313
Foster	4.517629	1.594968	2.83	0.005	1.39155	7.643708
mean(qual)						
Not Foster	8.276438	.1216513	68.03	0.000	8.038006	8.514871
Foster	8.209445	.1204224	68.17	0.000	7.973422	8.445469
mean(tcr)						
Not Foster	65.61905	.957413	68.54	0.000	63.74255	67.49554
Foster	63.8022	1.069522	59.65	0.000	61.70597	65.89842

var(e.bk)						
Not Foster	6.29192	.9752377			4.643531	8.525466
Foster	5.238666	.7798614			3.91296	7.01352
var(e.b1)						
Not Foster	3.095718	.4797206			2.284846	4.194361
Foster	4.327884	.6441391			3.232863	5.793805
var(qual)						
Not Foster	1.243119	.1918175			.9186911	1.682116
Foster	1.319642	.1956368			.9868829	1.764602
var(tcr)						
Not Foster	76.99773	11.88101			56.90294	104.1888
Foster	104.0927	15.43173			77.84486	139.1909
cov(qual,tcr)						
Not Foster	-.8320274	1.071323	-0.78	0.437	-2.931783	1.267728
Foster	-.3845261	1.229281	-0.31	0.754	-2.793873	2.024821

Note: [*] identifies parameter estimates constrained to be equal across groups.
LR test of model vs. saturated: chi2(4)   =   3.05, Prob > chi2 = 0.5489

These results show the separate estimates for qual $\rightarrow$ bk with all other paths constrained to be equal. Our first model had 0 degrees of freedom and hence had a chi-squared of 0. We would expect this model to do somewhat worse because we have constrained four of the five paths to be equal; however, our constrained model still fits virtually as well as the original model. We have a $\chi^2(4) = 3.05$, $p = 0.55$. This model provides an excellent fit to the data. We can run estat gof, stats(all) to obtain RMSEA $= 0.00$ and CFI $= 1.00$.

We can run the command estat eqgof (results shown below) to get the R^2 values for each endogenous variable, separately by group. We have more explanatory power to predict kindergarten behavioral problems among the foster home children, with $R^2 = 0.29$ for bk (behavior problems during kindergarten) compared with $R^2 = 0.17$ for nonfoster children. In the case of b1, behavioral problems at first grade, $R^2 = 0.23$ for nonfoster children and $R^2 = 0.19$ for foster children.

```
. estat eqgof
Group #1 (grp=0; N=84)

Equation-level goodness of fit
```

	fitted	Variance predicted	residual	R-squared	mc	mc2
observed						
bk	7.603526	1.311605	6.29192	.1724996	.4153307	.1724996
b1	4.021941	.9262232	3.095718	.2302926	.4798881	.2302926
overall				.1981286		

(depvars in first column)

```
mc  = correlation between depvar and its prediction
mc2 = mc^2 is the Bentler-Raykov squared multiple correlation coefficient
Group #2 (grp=1; N=91)

Equation-level goodness of fit
```

	fitted	Variance predicted	residual	R-squared	mc	mc2
observed						
bk	7.388829	2.150162	5.238666	.2910018	.5394458	.2910018
b1	5.318875	.9909908	4.327884	.1863159	.4316432	.1863159
overall				.3124117		

(depvars in first column)

```
mc  = correlation between depvar and its prediction
mc2 = mc^2 is the Bentler-Raykov squared multiple correlation coefficient
```

Should we run the `estat ginvariant` command on this solution? There is no reason to do so because our chi-squared value is not significant, hence, no change could have a significant improvement. However, if there were problems with model fit, we would want to run `estat ginvariant` to locate the problem. It would be possible to add equality constraints to the variances and the covariance, but we have no hypothesis about that, so we will allow them to be different.

Box 5.3. Using 0: and 1: is optional here but can be useful

We could have fit the model above by using this `sem` command:

```
. sem (bk <- qual tcr@a1)       ///
  (b1 <- qual@b1 tcr@c1 bk@d1), ///
  group(grp) cov(tcr*qual)
```

We did not need the prefix of 0: or 1:. When we do not include 0: and 1:, whatever constraints we specify—such as bk ← qual tcr@a1—apply to all groups. Sometimes it is very useful to include the group prefix as we did here.

Box 5.3. (*continued*)

Suppose you want to compare the effect of `qual` on `bk` with the effect of `tcr` on `bk`, assuming that both `qual` and `tcr` were measured on the same scale. Further, while constraining the paths to be equal in each group, we might not want to constrain them to be equal across groups. Our command would look like this:

```
. sem (0:bk<- qual@a1 tcr@a1) ///
(1:bk <- qual@b1 tcr@b1),    ///
group(grp) cov(tcr*qual)
```

Because both `qual` and `tcr` have the `a1` constraint in group 0:, these two parameter estimates are constrained to be equal. Because `qual` and `tcr` have the `b1` constraint in group 1:, these two parameter estimates are constrained to be equal. However, there is no cross-group equality constraint.

5.4.2   Fitting the model with the SEM Builder

We need to draw our model with constraints on all the parameter estimates we want to be the same in both groups; see figure 5.8.

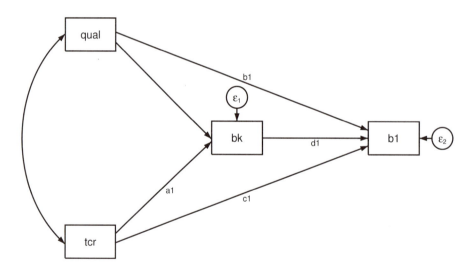

Figure 5.8. Model drawn showing parameters that are constrained

Now click on **Estimation** > **Estimate....** Click on **Group analysis** and enter `grp` as the *Grouping variable*. Leave the *Parameters that are equal across groups* blank; we indicated those in our drawing. Click on **OK** to get results shown on the figures. You could copy the figures as we did with the unconstrained results in figure 5.7.

5.4.3   A standardized solution

To present this to a general audience, it is often helpful to report a standardized solution based on the final model we fit. We simply replay the results using the command sem, standardized. When we constrain parameters to be equal for the unconstrained model, the standardized values will be different unless both groups have identical variances (see box 5.1 above).

```
. sem, standardized
Structural equation model                    Number of obs     =       175
Grouping variable  = grp                      Number of groups  =         2
Estimation method  = ml
Log likelihood     = -1675.1843
 ( 1)   [b1]0bn.grp#c.bk - [b1]1.grp#c.bk = 0
 ( 2)   [bk]0bn.grp#c.tcr - [bk]1.grp#c.tcr = 0
 ( 3)   [b1]0bn.grp#c.qual - [b1]1.grp#c.qual = 0
 ( 4)   [b1]0bn.grp#c.tcr - [b1]1.grp#c.tcr = 0
```

Standardized	Coef.	OIM Std. Err.	z	P>|z|	[95% Conf. Interval]	
Structural						
bk <-						
qual						
Not Foster	-.0857473	.0992319	-0.86	0.388	-.2802382	.1087435
Foster	.2144563	.0865502	2.48	0.013	.0448211	.3840915
tcr						
Not Foster	-.4137405	.0606671	-6.82	0.000	-.5326459	-.2948352
Foster	-.4879991	.0688388	-7.09	0.000	-.6229207	-.3530775
_cons						
Not Foster	4.438543	.8534955	5.20	0.000	2.765723	6.111364
Foster	2.455017	.8010072	3.06	0.002	.8850717	4.024962
b1 <-						
bk						
Not Foster	.3823779	.078314	4.88	0.000	.2288853	.5358704
Foster	.3277789	.0696216	4.71	0.000	.1913232	.4642347
qual						
Not Foster	-.0527564	.0709812	-0.74	0.457	-.1918771	.0863642
Foster	-.0472667	.0639099	-0.74	0.460	-.1725278	.0779944
tcr						
Not Foster	-.1685013	.0755852	-2.23	0.026	-.3166456	-.0203571
Foster	-.1703659	.0741106	-2.30	0.022	-.31562	-.0251117
_cons						
Not Foster	2.192219	.8041729	2.73	0.006	.6160687	3.768368
Foster	1.958848	.6852774	2.86	0.004	.6157288	3.301967
mean(qual)						
Not Foster	7.423131	.5830077	12.73	0.000	6.280456	8.565805
Foster	7.146374	.5399973	13.23	0.000	6.087999	8.204749
mean(tcr)						
Not Foster	7.478095	.5871739	12.74	0.000	6.327255	8.628934
Foster	6.253533	.4752482	13.16	0.000	5.322063	7.185002

var(e.bk)						
Not Foster	.8275004	.0505576			.7341124	.9327684
Foster	.7089982	.0735426			.578565	.8688367
var(e.b1)						
Not Foster	.7697074	.0663336			.6500832	.9113442
Foster	.8136841	.0536435			.7150542	.9259184
var(qual)						
Not Foster	1	.			.	.
Foster	1	.			.	.
var(tcr)						
Not Foster	1	.			.	.
Foster	1	.			.	.
cov(qual,tcr)						
Not Foster	-.0850437	.1083198	-0.79	0.432	-.2973467	.1272592
Foster	-.0328086	.1047156	-0.31	0.754	-.2380475	.1724303

LR test of model vs. saturated: chi2(4)   =   3.05, Prob > chi2 = 0.5489

5.4.4   Constructing tables for publications

It is important to present a summary table for multiple-group results. The table should show the unconstrained and the constrained results. It should also report both the unstandardized and standardized values along with measures of goodness of fit. Such a table for our path analysis model is shown in table 5.4. The error variances are not included in this table, but their inclusion may be expected in your research area. Read articles in your field that report results for structural equation models and adapt the table accordingly. Many journals would also like you to report a figure similar to figure 5.7 for your final model.

Table 5.4. Summary table for our multiple-group results

Relationship	Unconstrained solution				Constrained solution			
	Nonfoster child (N = 84)		Foster child (N = 91)		Nonfoster child (N = 84)		Foster child (N = 91)	
	B	β	B	β	B	β	B	β
Kindergarten behavior problems								
Presch quality	−0.19	−0.08	0.50*	0.21*	−0.21	−0.09	0.51*	0.21*
Presch teacher relationship	−0.10**	−0.33**	−0.15***	−0.54***	−0.13***	−0.41***	−0.13***	−0.49***
First grade behavior problems								
Presch quality	−0.09	−0.05	−0.10	−0.05	−0.10	−0.05	−0.10	−0.05
Presch teacher relationship	−0.06*	−0.24*	−0.02	−0.10	−0.04*	−0.17*	−0.04*	−0.17*
Kindergarten problems	0.30***	0.38***	0.27***	0.33**	0.28***	0.38***	0.28***	0.33***
R^2 Kindergarten behavioral problems		0.11		0.34		0.17		0.29
R^2 First grade behavioral problems		0.26		0.15		0.23		0.19
χ^2 by group	df = 0, 0.00, ns		df = 0, 0.00, ns		a		a	
χ^2 overall	df = 0, 0.00, ns				df = 4, 3.05, p = 0.55			
CFI	1.00				1.00			
RMSEA	0.00				0.00			

a Not reported because of constraints between groups.

* $p < 0.05$; ** $p < 0.01$; *** $p < 0.001$

5.5   Multiple-group comparisons of structural equation models

Multiple-group analysis can also be applied to a full structural equation model that includes a measurement model and a structural model. Figure 5.9 illustrates a fairly simple full structural equation model. We have multiple indicators for each of the latent variables in the structural model. A person's physical fitness, `Physical`, and attractiveness, `Appear`, are used to predict his or her relationship with peers, `Peerrel`. We have allowed two exogenous latent variables, `Appear` and `Physical`.

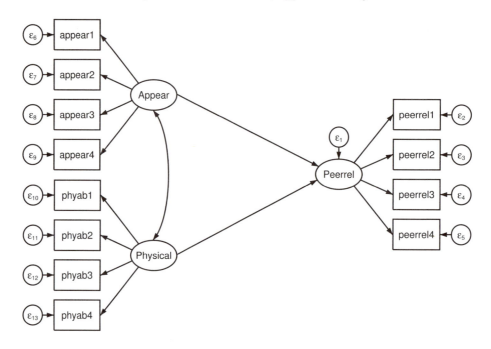

Figure 5.9. Full structural equation model predicting peer relations from a person's physical fitness and attractiveness

We will use data published by Marsh and Hocevar (1985). They provided summary data (means, standard deviations, and correlations). Stata used these summary statistics to illustrate how to create a dataset for structural equation modeling when you only have summary statistics and wish to perform multiple-group analysis (StataCorp 2013). Appendix B illustrates how you can do this when you have a single group. The dataset can be accessed by typing

```
. use http://www.stata-press.com/data/r13/sem_2fmmby.dta, clear
```

We are using two groups: a sample of fourth graders and a sample of fifth graders. We are interested in group differences in how well `Physical` and `Appear` predict `Peerrel`. It might be more interesting to have other groupings for this example to see whether there are differences between women and men or between preadolescents and adolescents. You can think of other groups we might want to compare. Adults in their early 20s might have a different emphasis on appearance and physical fitness than those in their 40s, for example, while those in their 40s might have a different emphasis than those in their 70s. Still, the fourth and fifth graders will be sufficient to illustrate how to do multiple-group structural equation modeling.

When analyzing summary data instead of raw data, we cannot use many common Stata commands such as `summarize` or `tabulate` to examine the data. We cannot use other Stata commands to assess normality. Importantly, we cannot use the `method(mlmv)` option to work with missing values. We can work with summary data based on listwise/casewise deletion. If, however, the summary data are based on pairwise deletion, then we may encounter estimation problems, because the correlation or covariance matrix cannot be inverted. Because the matrix may have a different subsample for each element, the resulting matrix may be impossible to obtain on any single sample.

To get an idea of the data we have, we can run `ssd describe`. Because Stata added notes to describe the dataset, we can also run the command `notes`. When you are ready to generate your own summary data, you will want to include a detailed set of notes so that a user of your data can know more about the original raw data you used.

```
. ssd describe
Summary statistics data from
http://www.stata-press.com/data/r13/sem_2fmmby.dta
  obs:           385              two-factor CFA
  vars:           16              25 May 2013 11:11
                                  (_dta has notes)
```

variable name	variable label
phyab1	Physical ability 1
phyab2	Physical ability 2
phyab3	Physical ability 3
phyab4	Physical ability 4
appear1	Appearance 1
appear2	Appearance 2
appear3	Appearance 3
appear4	Appearance 4
peerrel1	Relationship w/ peers 1
peerrel2	Relationship w/ peers 2
peerrel3	Relationship w/ peers 3
peerrel4	Relationship w/ peers 4
parrel1	Relationship w/ parent 1
parrel2	Relationship w/ parent 2
parrel3	Relationship w/ parent 3
parrel4	Relationship w/ parent 4

```
Group variable:  grade  (2 groups)
 Obs. by group:  134, 251
```

```
. notes
_dta:
    1.  Summary statistics data from Marsh, H. W. and Hocevar, D., 1985,
        "Application of confirmatory factor analysis to the study of
        self-concept: First- and higher order factor models and their invariance
        across groups", _Psychological Bulletin_, 97: 562-582.
    2.  Summary statistics based on 134 students in grade 4 and 251 students in
        grade 5 from Sydney, Australia.
    3.  Group 1 is grade 4, group 2 is grade 5.
    4.  Data collected using the Self-Description Questionnaire and includes
        sixteen subscales designed to measure nonacademic status: four intended
        to measure physical ability, four intended to measure physical
        appearance, four intended to measure relations with peers, and four
        intended to measure relations with parents.
```

We have 385 observations and 16 variables. At the bottom of the ssd describe
results, we see there are two groups and the grouping variable is grade. We also see
that we have 134 fourth graders and 251 fifth graders.

We could go directly to fitting the model, but it is best to have a satisfactory CFA
result before moving to the full structural equation model. If the CFA model in which
the latent variables are freely correlated rather than having structural paths does not fit
the data, then the full structural equation model is certain to fail. We would fit the CFA
model using the same steps we did at the start of this chapter to do multigroup CFA. We
will not repeat this here, but let us assume that we are satisfied with a three-factor CFA
with the four indicators of each of the latent variables as shown in figure 5.9. We also
assume that we have equal loadings across groups but have allowed the error variances
to differ. Now we are going to estimate the structural part of the model instead of
simply having the latent variables be correlated.

First, we will fit the model using two groups with the loadings and measurement
intercepts constrained to be equal across groups. This is the default: Stata assumes you
will let the variances of the error terms vary but requires the loadings to be equal to
ensure that the latent variables have the same meaning in both groups. Our command is

```
. sem (Peerrel -> peerrel1 peerrel2 peerrel3 peerrel4)      /// Measurement
>     (Appear -> appear1 appear2 appear3 appear4)            /// Measurement
>     (Physical -> phyab1 phyab2 phyab3 phyab4)              /// Measurement
>     (Appear Physical -> Peerrel), group(grade)            //  Structural
  (output omitted)
```

	Coef.	OIM Std. Err.	z	P>|z|	[95% Conf. Interval]	
Structural						
Peerrel <-						
Appear						
1	.4916053	.0755888	6.50	0.000	.343454	.6397566
2	.3491358	.0550062	6.35	0.000	.2413256	.4569459
Physical						
1	.2770591	.1094364	2.53	0.011	.0625677	.4915504
2	.1890172	.0699383	2.70	0.007	.0519407	.3260937

Measurement						
peerrel1 <-						
Peerrel						
[*]	1	(constrained)				
_cons						
[*]	8.724141	.1184925	73.63	0.000	8.4919	8.956382
peerrel2 <-						
Peerrel						
[*]	1.205956	.1047847	11.51	0.000	1.000582	1.41133
_cons						
[*]	7.874054	.1361587	57.83	0.000	7.607187	8.14092
peerrel3 <-						
Peerrel						
[*]	1.438179	.1177689	12.21	0.000	1.207356	1.669002
_cons						
[*]	7.418234	.1555249	47.70	0.000	7.113411	7.723057
peerrel4 <-						
Peerrel						
[*]	1.2616	.1033987	12.20	0.000	1.058942	1.464258
_cons						
[*]	8.19665	.1376685	59.54	0.000	7.926825	8.466475
appear1 <-						
Appear						
[*]	1	(constrained)				
_cons						
[*]	7.517397	.1900011	39.57	0.000	7.145002	7.889792
appear2 <-						
Appear						
[*]	1.099445	.070083	15.69	0.000	.9620852	1.236806
_cons						
[*]	7.16282	.2108585	33.97	0.000	6.749545	7.576095
appear3 <-						
Appear						
[*]	1.191424	.0748902	15.91	0.000	1.044642	1.338207
_cons						
[*]	7.200432	.219001	32.88	0.000	6.771198	7.629666
appear4 <-						
Appear						
[*]	1.031239	.0694995	14.84	0.000	.8950221	1.167455
_cons						
[*]	7.370821	.1949526	37.81	0.000	6.988721	7.752921
phyab1 <-						
Physical						
[*]	1	(constrained)				
_cons						
[*]	8.334832	.1356629	61.44	0.000	8.068937	8.600726

```
phyab2 <-
   Physical
       [*]      .8518032    .0882102     9.66    0.000      .6789144    1.024692
     _cons
       [*]      8.344266    .1253379    66.57    0.000      8.098609    8.589924
```

```
phyab3 <-
   Physical
       [*]      1.380531    .1070781    12.89    0.000      1.170662      1.5904
     _cons
       [*]      8.347134    .1722181    48.47    0.000      8.009593    8.684675
```

```
phyab4 <-
   Physical
       [*]      1.226901    .0946628    12.96    0.000      1.041365    1.412437
     _cons
       [*]      8.612711    .1533223    56.17    0.000      8.312204    8.913217
```

```
mean(Appear)
        1             0    (constrained)
        2     -.0377243    .2151596    -0.18    0.861     -.4594293    .3839807
mean(Physical)
        1             0    (constrained)
        2      -.090445    .1472553    -0.61    0.539       -.37906      .19817
```

```
var(e.peerr~1)
   (output omitted)
```

```
cov(Appear,
  Physical)
        1      1.473904    .2889497     5.10    0.000      .9075727    2.040235
        2      .9633147    .1825246     5.28    0.000      .6055729    1.321056
```

Note: [*] identifies parameter estimates constrained to be equal across groups.
LR test of model vs. saturated: chi2(121) =   267.19, Prob > chi2 = 0.0000

. estat gof, stats(all)

   (output omitted)

. estat ggof

   (output omitted)

. estat eqgof

   (output omitted)

From the estat gof postestimation command (results not shown), we learn that $\chi^2(121) = 267.19$, RMSEA = 0.08, and CFI = 0.93. From the estat eqgof postestimation command, we learn that R^2 for our Peerrel equation in the first group, fourth graders, is 0.73 and in the second group, fifth graders, is 0.45. The difference in R^2 values suggests that more factors go into explaining peer relations among the older students than just attractiveness and physical fitness.

For the fourth graders, the sem results show that the unstandardized path coefficient for Appear → Peerrel is 0.49, and for fifth graders it is smaller at 0.35. For Physical → Peerrel, the unstandardized path coefficient is 0.28 for fourth graders compared with 0.19 for fifth graders. We might ask for a standardized solution, running `sem, standardized` (results not shown). These results are $\beta = 0.68$ for `Appear` $\rightarrow$ `Peerrel` among fourth graders and $\beta = 0.55$ among fifth graders. The results are $\beta = 0.25$ for `Physical` $\rightarrow$ `Peerrel` among fourth graders and $\beta = 0.21$ among fifth graders.

Let us rerun the model with both of the structural path coefficients constrained across groups (results are not shown for the following set of commands).

```
. sem (Peerrel -> peerrel1 peerrel2 peerrel3 peerrel4)     ///
      (Appear -> appear1 appear2 appear3 appear4)          ///
      (Physical -> phyab1 phyab2 phyab3 phyab4)            ///
      (Appear@a1 Physical@b1 -> Peerrel), group(grade)     //   a1/b1 constraints
. estat gof, stats(all)
. estat ggof
. estat eqgof
```

The chi-squared for this model is $\chi^2(123) = 275.09$, $p < 0.001$, and RMSEA $= 0.08$, CFI $= 0.93$. How does this compare with our initial model in which the structural paths were not constrained? The difference in chi-squared is 7.90, and the difference in degrees of freedom is 2; thus our chi-squared test for the difference is $\chi^2(2) = 7.90$, $p < 0.05$.

Now we need to make a decision. This model does not do much worse in terms of the RMSEA and CFI, but in terms of chi-squared, it does significantly worse. With an extremely large sample, a chi-squared will be significant even when the inequalities are trivial, but our sample is not extremely large. Many researchers would probably say that the equality constraints do significantly worse and conclude that the structural paths are different for fourth graders than they are for fifth graders.

The next question is whether both paths are different or just one of them is different. We can explore this question by using the `estat ginvariant` command. Here are our results:

```
. estat ginvariant
```
Tests for group invariance of parameters

	Wald Test			Score Test		
	chi2	df	p>chi2	chi2	df	p>chi2
Structural						
Peerrel <-						
Appear	.	.	.	6.823	1	0.0090
Physical	.	.	.	4.730	1	0.0296
Measurement						
peerr~1 <-						
Peerrel	.	.	.	3.144	1	0.0762
_cons	.	.	.	1.119	1	0.2901
peerr~2 <-						
Peerrel	.	.	.	2.953	1	0.0857
_cons	.	.	.	0.156	1	0.6933

peerr~3 <-						
Peerrel	.	.	.	0.507	1	0.4766
_cons	.	.	.	1.072	1	0.3005
peerr~4 <-						
Peerrel	.	.	.	0.073	1	0.7875
_cons	.	.	.	0.989	1	0.3201
appear1 <-						
Appear	.	.	.	2.744	1	0.0976
_cons	.	.	.	2.066	1	0.1506
appear2 <-						
Appear	.	.	.	0.471	1	0.4924
_cons	.	.	.	4.227	1	0.0398
appear3 <-						
Appear	.	.	.	1.075	1	0.2999
_cons	.	.	.	0.047	1	0.8293
appear4 <-						
Appear	.	.	.	0.343	1	0.5579
_cons	.	.	.	1.579	1	0.2089
phyab1 <-						
Physical	.	.	.	2.093	1	0.1480
_cons	.	.	.	0.804	1	0.3699
phyab2 <-						
Physical	.	.	.	0.115	1	0.7340
_cons	.	.	.	0.380	1	0.5378
phyab3 <-						
Physical	.	.	.	1.711	1	0.1909
_cons	.	.	.	2.963	1	0.0852
phyab4 <-						
Physical	.	.	.	0.042	1	0.8375
_cons	.	.	.	4.309	1	0.0379
var(e.peer~1)	0.375	1	0.5405	.	.	.
var(e.peer~2)	0.231	1	0.6305	.	.	.
var(e.peer~3)	0.079	1	0.7785	.	.	.
var(e.peer~4)	0.374	1	0.5411	.	.	.
var(e.appe~1)	0.461	1	0.4971	.	.	.
var(e.appe~2)	0.901	1	0.3424	.	.	.
var(e.appe~3)	1.776	1	0.1826	.	.	.
var(e.appe~4)	0.485	1	0.4863	.	.	.
var(e.phyab1)	0.711	1	0.3993	.	.	.
var(e.phyab2)	3.203	1	0.0735	.	.	.
var(e.phyab3)	0.469	1	0.4935	.	.	.
var(e.phyab4)	0.548	1	0.4593	.	.	.
var(e.Peer~1)	0.065	1	0.7994	.	.	.
var(Appear)	2.034	1	0.1538	.	.	.
var(Physical)	0.041	1	0.8402	.	.	.

cov(Appear, Physical)	3.360	1	0.0668	.	.	.

The results may be a bit confusing. The columns labeled `Wald Test` tell us what would happen if we constrained parameters to be equal that are currently free to differ. For example, the variance of the error term for `peerrel1`, `e.peerrel1`, is not constrained to be equal across groups. If we put an equality constraint on it, `var(e.peerrel1@a)`, the model would not fit significantly worse: Wald $\chi^2(1) = 0.38$, $p = 0.54$. Although we have not put any equality constraints on the error variances, these results suggest we might have tried a more restrictive model that did place equality constraints on all of them.

There are two limitations to interpreting these Wald chi-squared values. First, these will only estimate the reduction in our likelihood-ratio chi-squared if we constrained a parameter to be equal across groups. Second, these are for each parameter estimate individually. Although an equality constraint on none of the error variances individually would have a significant effect, these results do not tell us what happens with a combination of them.

The last three columns are labeled `Score Tests` (Lagrange multiplier tests). These indicate how much our chi-squared test would be reduced if we removed an equality constraint. You will remember that Stata, by default, put an equality constraint on each of the loadings. Removing the constraint on any individual loading would not result in a significant improvement of chi-squared; that is, they are all less than $\chi^2(1) = 3.84$, $p < 0.05$.

However, if we look at the chi-squared for the structural paths, we see that removing the constraint on either of them would significantly improve our fit. We could use this information in combination with the difference between the model that constrains neither and the model that constrains both to support a model that constrains neither. We can see that both structural paths, considered one at a time, are statistically significantly different across groups. Because our model that constrained both of these to be equal did significantly worse than our model that did not constrain them, we might decide to have both of them be different and go forward with the unconstrained initial model.

Figure 5.10 is one way we could present our results. To simplify the figure, I leave off the error variances. I do include the unstandardized loadings which, by default, are equal for both groups. We report both the unstandardized and standardized structural paths for the two groups. It would appear that this model fits the data at least adequately. Appearance and physical attractiveness are both significant predictors of peer relations, and this is true for both fourth and fifth graders.

Both predictors are somewhat stronger for fourth graders than they are for fifth graders. The $\beta = 0.68$ for the standardized effect of `Appear` $\rightarrow$ `Peerrel` for fourth graders compared with $\beta = 0.55$ for fifth graders. The $\beta = 0.25$ for the standardized effect of Physical → Peerrel for fourth graders compared with $\beta = 0.21$ for fifth graders. Although the coefficients are only slightly weaker for fifth graders, the model fits significantly worse when the unstandardized efficients are constrained to be equal. The differences in the effects is reflected in a larger R^2 for fourth graders ($R^2 = 0.69$) compared with fifth graders ($R^2 = 0.48$).

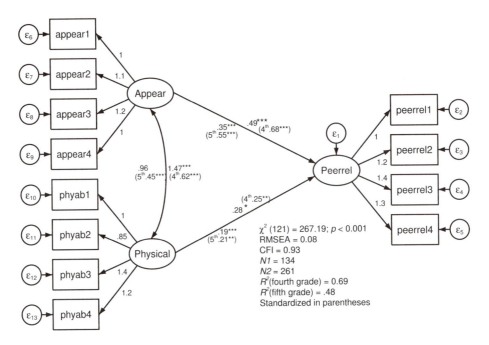

Figure 5.10. Multigroup comparison of fourth and fifth grade students ($**p < 0.01$; $***p < 0.001$)

5.6   Exercises

Exercises 1–7 are based on the summary statistics dataset sem_2fmmby.dta. You may access this dataset by typing

```
. use http://www.stata-press.com/data/r13/sem_2fmmby.dta, clear
```

Each of these exercises involves some type of comparison of fourth and fifth grade children. You should interpret the results you obtain for each exercise.

1. Describe the dataset and any notes attached to it.

2. Fit a multiple-group CFA for peer relations, parent–child relations, physical fitness, and appearance with grade as your grouping variable.

3. Examine the modification indices of what you did in exercise 2. Why might you correlate the errors for appear1 and appear2? Estimate the CFA with these errors correlated.

4. Fit a model with equal loadings and keep the correlated error you had for exercise 3. Why is it important to have equal loadings?

5. Fit the model with equal loadings, the correlated error, and equal error variances. Can we treat the error variances as equal?

6. Fit the model with equal loadings, the correlated error, and equal intercepts.

7. Fit the model with equal loadings, the correlated error, and equal intercepts, but allow the means on the latent variables to differ.

8. Use the hsb2artificial.dta dataset. These data involve the relative importance of reading and writing ability to predict math ability as well as the relative importance of reading, writing, and math ability to predict science ability. Draw a figure to represent this model.

    a. Fit the model with all paths allowing for group differences where socioeconomic status is your grouping variable.

    b. Fit the model with all structural paths constrained to be equal. Compare with your results from part a of this exercise.

6 Epilogue–what now?

You have seen several specific applications of structural equation modeling. These examples were selected to represent as broad a range of applications as possible. You may find an example that is exactly what works for your own research, or you may need to mix features of several of these examples to meet your needs. These techniques greatly expand the types of questions we can address in our research; as you worked through the examples, you probably thought of several studies you could do yourself.

We gave some attention to interpretation, but I did not include detailed interpretation for each of the examples. There have been many attempts to do this; an article by Schreiber et al. (2006) is both highly accessible and quite comprehensive. Having worked your way through this book, you will find this article to be a great next step. You are prepared to fully digest its content. Schreiber et al. go into great detail on what to include in a confirmatory factor analysis or structural equation modeling article. They go into evaluating how well your data meet assumptions in more detail than we have done here. They go over several measures of goodness of fit and discuss criteria. And they provide considerable explanation on what statistical information to report and how to report that information in both figures and tables.

I designed this book to meet the needs of two types of researchers. The primary audience is people who are new to structural equation modeling and who will read this book and work their way through the examples and exercises to facilitate learning. The secondary audience is people who already know structural equation modeling and just want to see how to estimate models using Stata. Whichever type of reader you are, keep this book close at hand and use it as a reference.

The Stata command structure is much more straightforward than the command structure of other programs. Most other structural equation modeling software can do graphs, but Stata graphs are truly production quality. If you rely on the command structure to estimate your model, you will at least use the SEM Builder to create your final graphs. The commands that Stata uses are a balance of parsimony and readability. Some other programs have extremely verbose command structures and are so verbose that they are difficult to understand. Other programs are extremely cryptic in their command structure, and this also makes them difficult to understand. At this point, I hope that you can look at any of Stata's structural equation modeling commands and visualize the underlying figure, and that you can look at a figure and easily write the commands to estimate it.

6.1   What is next?

Extensions to SEM, known in Stata as generalized structural equation modeling (GSEM), were introduced in Stata 13. GSEM allows for fitting models with a variety of types of outcomes by specifying the family and link for the outcome variable. It also allows for fitting multilevel models. Generalized structural equation models can be fit using the `gsem` command or using the SEM Builder.

When discussing the SEM Builder, we have not used the ⬛, ⬜, ⬭, and ⬛ icons. These are used when fitting generalized structural equation models. Models with generalized response variables are fit using the rectangle with bars along the top and bottom showing the distributional family and link function of the named outcome variable. Here is a listing of the families and links available at this writing.

```
default    family(gausian) link(identity)
cloglog    family(bernoulli) link(cloglog)
gamma      family(gamma) link(log)
logit      family(bernoulli) link(logit)
nbreg      family(nbreg mean) link(log)
mlogit     family(multinomial) link(logit)
ocloglog   family(ordinal) link(cloglog)
ologit     family(ordinal) link(logit)
oprobit    family(ordinal) link(probit)
poisson    family(poisson) link(log)
probit     family(bernoulli) link(probit)
           family(gaussian) link(log)
           family(binomial) link(logit)
           family(binomial) link(probit)
           family(binomial) link(cloglog)
```

Although these capabilities are beyond the scope of this book, they are covered in the Structural Equation Modeling reference manual, and numerous examples are shown. Depending on your experience with these estimators outside of SEM, this book should have prepared you to explore their applications within SEM. These estimators allow extensions of SEM to cover models where the outcome variables may be binary (`logit`, `probit`, `cloglog`), categorical (`mlogit`), ordinal (`ologit`, `oprobit`, `ocloglog`), count (`poisson`, `nbreg`), or positive continuous but skewed (`gamma`).

For example, if you wanted to run a logistic regression of whether a person graduates from college using his or her gender, adoption status, and mother's education status as predictors, you would represent the model as

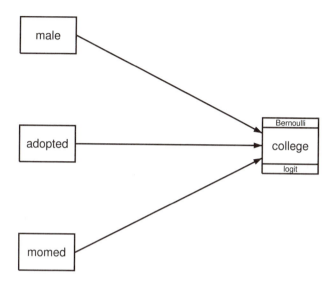

You could fit the model using **gsem** with a syntax similar to **sem** syntax but specifying the family and link.

```
gsem (college <- make adopted momed), family(bernoulli) link(logit)
```

An equally important extension of SEM is that GSEM can estimate multilevel models. You might be interested in how much customers like a new item on a menu at a chain of fast food outlets. You think gender, education, and age are important predictors. For example, you might think that younger women with high educations will like the item more than other groups. You have a sample of 100 people who try the new menu item in each of 40 outlets. Thus you have 4,000 participants in your sample. However, your 4,000 people are not independent, because the people sampled at any one outlet might be similar to each other. In other words, there is variance between outlets as well as variance between people, within outlets. Multilevel analysis allows you to handle this type of problem by incorporating the between-outlet variance. In the Builder, the double oval icon allows us to include latent variables at the outlet level and perform a multilevel analysis. Using the **gsem** command, we can fit multilevel models by specifying the level of a latent variable in square brackets. In this example, we might include a latent variable named L[outlet], where L is the name of a latent variable that varies at the outlet level, and outlet is the variable in your dataset identifying the outlet where the customer sampled the new menu item.

Another example requiring multilevel analysis would be a study of physician satisfaction where the physicians might be nested in centers and centers might be nested in larger organizations. Education research often needs to do multilevel analysis because children are nested in classrooms, and the classrooms are nested in schools. Stata's GSEM can estimate multilevel models with any degree of nesting. This is a valuable extension.

GSEM features in Stata offer many other extensions to the capabilities we have covered in this book. Some of these include Rasch and item response theory modeling, random coefficients modeling, extensions of the Heckman selection model, and endogenous treatment-effects models. Your next step may be learning the extensions that are specific to your research applications.

A The graphical user interface

A.1 Introduction

Stata's graphical user interface, the SEM Builder, is more than a simple drawing program. It is designed specifically for drawing and fitting structural equation models. The Builder can be used to draw any model covered in this book. Stata can interpret the drawing that you create and write the code to fit the model. Then Stata can send the results back to the SEM Builder so that parameter estimates appear in your drawing. You have a lot of flexibility in what results are included in your final version, and you can also add explanatory notes to the figure. Beyond this, the postestimation commands for **sem** can be run without leaving the SEM Builder. The drawing you create, including the estimates, is production quality, ready for publication.

In this appendix, you will learn how to use the SEM Builder based on Stata 13. It is good to work through this appendix before trying to use the Builder yourself. Though we used the Builder in the text, I did not present a systematic introduction to its capabilities—that is the purpose of this appendix. If you work through this appendix, you will be able to create all the figures that appear in this book.

As you work through the following pages, do not worry if you do not understand the statistical information presented here; it is fully explained in the main text. Here I make no attempt to explain the statistical information.

The best way to learn how to draw a model with the SEM Builder is to use the SEM Builder, so open Stata and let us get started.[1] We will use with a dataset that is an example from Stata's *Structural Equation Modeling Reference Manual*. You can open this dataset by typing

```
. use http://www.stata-press.com/data/r13/sem_sm2.dta
(Structural model with measurement component)
```

---

1. Use caution as you learn how to use the SEM Builder because you will likely change some of the defaults. These changes stick in the SEM Builder and may not be what you want in a subsequent model you are drawing. To restore Stata defaults for the SEM Builder, click on **Settings** > **Settings Defaults** when you open your next SEM Builder window before drawing your next diagram. If you use a Mac, be sure to read box 1.A in the appendix to chapter 1 to learn where to access your **Settings** menu; information on the differences in operating systems is also presented in this appendix section A.2.

These data are in summary format rather than being raw data.[2] We can run the command `ssd describe` to obtain a description of the data:

```
. ssd describe
Summary statistics data from
http://www.stata-press.com/data/r13/sem_sm2.dta
    obs:            932              Structural model with measurem..
   vars:             13              25 May 2013 11:45
                                     (_dta has notes)

variable name                        variable label

educ66                               Education, 1966
occstat66                            Occupational status, 1966
anomia66                             Anomia, 1966
pwless66                             Powerlessness, 1966
socdist66                            Latin American social distance, 1966
occstat67                            Occupational status, 1967
anomia67                             Anomia, 1967
pwless67                             Powerlessness, 1967
socdist67                            Latin American social distance, 1967
occstat71                            Occupational status, 1971
anomia71                             Anomia, 1971
pwless71                             Powerlessness, 1971
socdist71                            Latin American social distance, 1971
```

We will build a model and estimate the parameters.

Open up the SEM Builder in Stata. You can do this either by typing `sembuilder` in the Command window or by using the menu system—**Statistics > SEM (structural equation modeling) > Model building and estimation**.

A.2   Menus for Windows, Unix, and Mac

Windows, Unix, and Mac computers have identical capabilities but lay out their menus differently. Each operating system has a set of menu items available from a bar across the top of the SEM Builder, but you access them in a slightly different way. On Mac computers, you access some of the menus by first clicking on the **Tools** button on the right side of the toolbar. If you use a Mac but cannot see this icon, your window may be too small; drag the SEM Builder window wider until the **Tools** button appears.

A.2.1   The menus, explained

For Windows and Unix, the SEM Builder has a bar across the top with straightforward names for each menu (**File**, **Edit**, **Object**, **Estimation**, **Settings**, **View**, and **Help**); you can see this bar at the top of figure A.2. Regardless of your operating system, take a minute to experiment with the menus to get an idea of the many special features that are available. For example, in the **Object** drop-down menu, there is an option to **Align**

---

2. How to enter data in summary format is described in appendix B.

objects. After you draw parts of your model, you can highlight them and then use the alignment feature to make them line up properly.

The **Estimation** drop-down menu has your statistical capabilities. You can **Estimate...** your model here or **Clear Estimates** if you want to make changes in your model. There are additional options for **Testing and CIs**, measures of **Goodness of fit**, **Group statistics** for comparing groups (such as women and men), and **Predictions** for predicting things like factor scores, and there is an **Other** category with more specialized postestimation commands.

The **Settings** drop-down menu is very useful: it lets you make adjustments in how your variables, connections (paths), and text are displayed. Your adjustments can apply to all latent variables, just endogenous variables, or other groups of variables that you specify. Here is where you will control what estimates appear in your figure and how they are displayed. The default display is excellent, but the **Settings** can be used to fine-tune the display if that is what you want.

The **View** drop-down menu has options for zooming in or out. This helps when you are working on a complex model where fitting everything in the right place is challenging. You can also expand your window for the SEM Builder to fill your screen and then click on the **Fit in Window** button. I find this extremely useful because it is easier to construct a model when you have the largest possible workspace for doing this. You can also **Show Grid** lines here. These will not appear when you print your graph, but they are extremely helpful when you are trying to arrange objects. When you fit a model in the unstandardized metric but want to see standardized results, you simply click on **View** > **Standardized Estimates**—you often want these.

A.2.2   The vertical drawing toolbar

On the left side of the Builder, you will see drawing tools (see figure A.1; these are the same for Windows, Unix, and Mac). From top to bottom, these are the **Change to generalized SEM** button, the **Select** tool, the **Add Observed Variable** tool, the **Add Generalized Response Variable** tool, the **Add Latent Variable** tool, the **Add Multilevel Latent Variable** tool, the **Add Path** tool, the **Add Covariance** tool, the **Add Measurement Component** tool, the **Add Observed Variables Set** tool, the **Add Latent Variables Set** tool, the **Add Regression Component** tool, the **Add Text** tool, and the **Add Area** tool. The names of the tools are fairly self-explanatory as to what they do. The **Select** tool is used to click or double-click on objects that you want to modify. Sometimes the **Add Measurement Component** tool is the easiest way to create a latent variable that has multiple indicators. I described how to use this tool in the appendix to chapter 1. The **Change to generalized SEM** button, the **Add Generalized Response Variable** tool, and the **Add Multilevel Latent Variable** tool are used for fitting multilevel SEMs or SEMs with generalized-response variables, which are beyond the scope of this book.

Figure A.1. The drawing toolbar

A.3   Designing a structural equation model

We will begin with a finished product of sufficient quality that you could print it in a journal article. Figure A.2 represents a structural equation model and was created using the SEM Builder in Windows. We have constructed this model with labels identifying the objects. We have three latent variables: `Alien67`, `Alien71`, and `SES66`.

The idea is that your SES66 (measured in 1966) influences your alienation in 1967 and in 1971: SES66 → Alien67 and SES66 → Alien71. Additionally, your alienation in 1967 influences your alienation in 1971: Alien67 → Alien71. SES66 is a latent exogenous variable (all arrows point away from an exogenous variable). Both Alien67 and Alien71 are latent endogenous variables (one or more arrows point toward an endogenous variable).

We have two indicators of SES66: educ66 and occstat66. Both of these are called observed endogenous variables. We have a pair of indicators for Alien67, anomia67 and pwless67, and a pair of indicators for Alien71, anomia71 and pwless71. These indicators are also called observed endogenous variables.

We have two kinds of error variables. We have errors in the latent endogenous variables: ϵ_5 and ϵ_8. The variances of these errors are the unexplained portion of the variances in these latent endogenous variables. We also have measurement errors: ϵ_1, ϵ_2, ϵ_3, ϵ_4, ϵ_6, and ϵ_7, which are errors in the observed variables.

Rectangles are used to enclose each of our observed variables (educ66, occstat66, anomia67, pwless67, anomia71, and pwless71). Ovals are used to enclose each of our latent variables (SES66, Alien67, and Alien71) as well as each of our error variables (ϵ_1– ϵ_8). The proper direction of the arrows is critical.

Why do some variables start with a capital letter? Although it is not necessary in the SEM Builder, I follow the convention for the sem command by starting all latent variables with a capital letter, for example, Alien67. To distinguish latent variables from their observed indicators, I use only lowercase letters for observed indicators,[3] for example, anomia67. If you have a dataset that puts variable names for all variables or some variables in capital letters or has a mix of capital and lowercase letters, you will want to change all observed variable names to be lowercase. A single Stata command can accomplish this: rename (*), lower. This goes through each variable name in your dataset and changes it to be lowercase.

The objects in this model are labeled with letters. Notice that figure A.2 does not have an observed exogenous variable (E). An example of such a variable might be gender[4] if you wanted to add a path from gender to Alien67 (gender → Alien67) and gender to Alien71 (gender → Alien71). Selected variables have two attached letters. If we wanted to make a change across all latent variables, such as increasing the size of the ovals to accommodate longer names, we could easily refer to all of them (A). However, we could make a change only to the oval around SES66 by just referring to latent exogenous variables (C). The final point to mention is that the paths Alien67 → pwless67 and Align71 → pwless71 are constrained to be equal.

---

3. I make only one exception to this rule: when I have interaction terms, I use a capital X between the component names; for example, I might use the name genXinc to refer to the interaction of the variables gender and income.

4. Notice that gender is all lowercase because it is an observed rather than a latent variable.

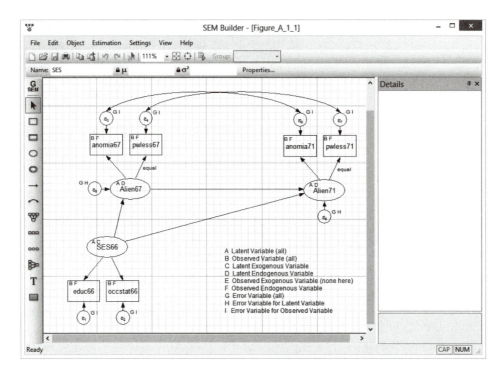

Figure A.2. A model built in the SEM Builder with types of objects labeled

We begin by working backward. Our final model will look like figure A.3, including all the parameter estimates.

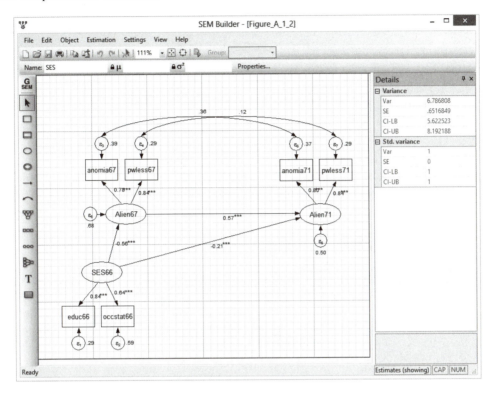

Figure A.3. Final figure

Box A.1. Draw a figure freehand

Until you are very familiar with the SEM Builder, I suggest that you first draw a freehand sketch of your model. This way, you have the opportunity to decide the best way to arrange the variables. Usually we work from the left to the right. Perhaps you want to have SES66 farther to the left than it is in our final model. You also might like to flip the figure so that SES66 is above Alien66 and Alien67. You also need to decide how close things are to each other and where you want error terms located. I suggest you sketch this out with paper and pencil first because it is easier to draw another quick sketch than it is to move objects around in the SEM Builder.

A.4   Drawing an SEM model

If you have not already done so, open Stata and open the dataset with the following command:

```
. use http://www.stata-press.com/data/r13/sem_sm2.dta, clear
```

Also open the SEM Builder if you have not yet. You can do this either by running the command `sembuilder` or by using the drop-down menu **Statistics > SEM (structural equation modeling) > Model building and estimation**. Expand your window and then use the **Fit to Window** button or change the **Zoom** percentage to make your workspace for the figure as large as possible.

We are now ready to add the variables to our model. It would be easiest to add each latent variable and its corresponding indicators using the **Add Measurement Component** tool. However, to demonstrate some features of the SEM Builder, we will instead add each component of this model to the diagram separately.

To begin, click on the **Add Latent Variable** tool (the oval) in the drawing toolbar and then click on your workspace where you want SES66; repeat this for `Alien67` and `Alien71`. Click on the **Select** tool at the top of the drawing toolbar, and click on the oval you want to label `Alien67`. When you click on a latent variable oval, the SEM Builder places a blue box around it, and a toolbar appears just above the drawing area. In the new toolbar, type in the box next to **Name** the name of the latent variable, `Alien67`. Repeat this procedure to name each of the latent variables. At this point, you have something similar to figure A.4.[5]

---

5. Figure A.4 shows the Properties window. To view this window, click on **View > Show Property Sheet**, and to hide it click on **View > Hide Property Sheet**. When drawing more complex models, you may want to hide this window to maximize the space you have for your drawing.

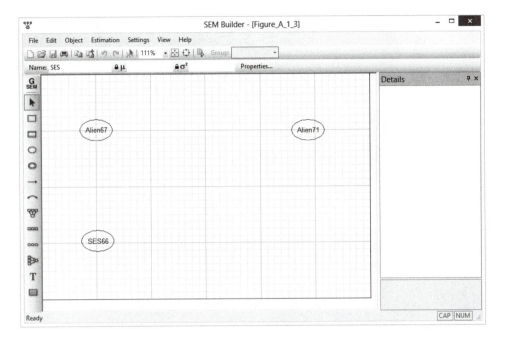

Figure A.4. Laying out our latent variables

Now you can move the ovals around to where you want them. Make sure that there is good alignment for `Alien67` and `Alien71`: Click on the **Select** tool, and then click and drag over these two latent variables. Once they are both highlighted, choose the **Object** > **Align** menu item and click on the **Horizontal Bottom** option.

Now we want to add rectangles for our observed variables. Click on the **Add Observed Variable** tool (the rectangle) in the drawing toolbar. Just like with the latent variable ovals, you now click on the diagram to create a rectangle for each of our six observed variables (two per latent variable). You can move the rectangles around and use **Object** > **Align** to get correct alignment. Make sure you leave enough room for the error terms; this is especially important for the indicators of `Alien67` and `Alien71` because you will need room for the curved lines representing correlated error terms.

When you click on the **Select** tool and then click on one of the rectangles, you can choose a variable name from the *Variable* drop-down menu that becomes available in the top toolbar. You can resize the rectangles if your actual variable names are too long to fit, but you may prefer to rename your variables to have just eight or fewer characters before starting to use the SEM Builder. Your model may look something like figure A.5 now.

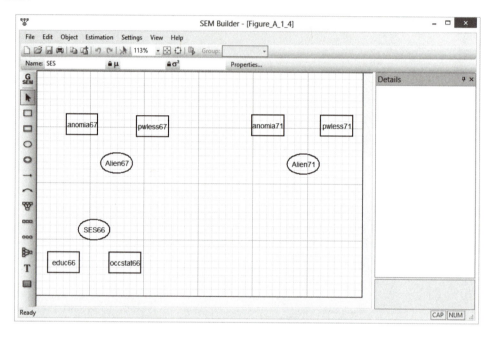

Figure A.5. Adding observed indicators

We are now ready to put in the paths. Click on the **Add Path** tool (the thin arrow pointing to the right) in the drawing toolbar. Click near the source and drag to the destination to draw the path. The objects are green at both ends of the path when you have made the connection.

When you insert a path, Stata inserts the corresponding error terms. The two observed endogenous variables, `educ66` and `occstat66`, have their error terms, but notice there is no error term for **SES66**. This is an exogenous latent variable, and so we are not trying to explain it—it does not need an error term. Do not worry if the error terms are in a different order than what you see in figures A.2 and A.3. The subscripts on the error terms will vary depending on the order in which you input the variables. This does not matter. Your model may now look like figure A.6.

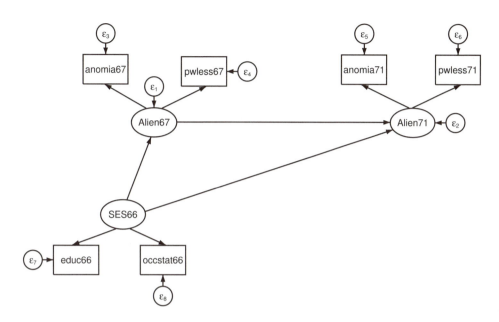

Figure A.6. Adding paths

So you think this looks ugly. You really do not like the location of some of the error terms. You might also like to have the arrows from the latent variables go to the centers of the observed variables.

Click on the **Select** tool, and then click on `educ66`. On the toolbar that appears just above the figure, click on the ⟲ button to bring the error term to a 6 o'clock position below `educ66`. You might want to make similar changes for other error terms that are not nicely located in figure A.6. This does what we need, but sometimes you may prefer to fine-tune the location of the error terms by clicking on them and dragging them to where you want them.

Next click on any of the paths, and notice the two small circles that appear at the ends of the path. Click on one of these circles and drag the path end to where it belongs, and then click on the other circle and do the same. You can do this to fine-tune where the direct paths intersect the objects.

Let us also move the `SES66` latent variable and its indicators up and to the right a bit. Using the **Select** tool, click and drag over `SES66` and its indicators, including the error terms, to highlight the block of variables. You can now click and drag to move all the highlighted objects as a block.

Sometimes we need to correlate error terms. Let us correlate the error terms for `anomia67` and `anomia71` and the error terms for `pwless67` and `pwless71`. Click on the

Add Covariance tool (the arched arrow) on the drawing toolbar, and draw in the lines. When you do this, you may find that the curves are quite pronounced; perhaps they even extend above the top of your figure. Click on a curve to see a light blue line with a circle at the top appear at one end of the curve. Click on the circle and drag to reshape your curve. After you add the correlated error paths, your model looks like figure A.7.

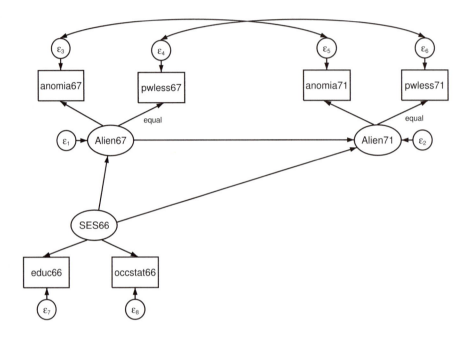

Figure A.7. Final figure

This looks a lot better. For reasons explained in the main text, we also might want to put equality constraints on the path from `Alien67` to `pwless67` and `Alien71` to `pwless71`. We do this by assigning the same label to each of these paths. In figure A.7, I used the label `equal` for both. To do this, click on the **Select** tool, and then click on one of the paths we want to constrain. In the top toolbar, a **aβ** box will appear where you will type `equal`. Then click on the other path to constrain and do the same. Because both paths have the same attached label, Stata will constrain them to be equal for the unstandardized solution.

We are done! Let us copy this figure into a Word document. First, resize the canvas so that no more than half an inch of space is between the edge of the figure and the edge of the canvas on all sides. The easiest way to accomplish this is to **Select** the entire drawing and move it to the lower left corner of your workspace; now when you resize the canvas, everything will fit nicely.

Once the canvas is the right size, again **Select** your entire figure. You can use *Ctrl+c* to copy the figure. Move to a Word document, and use *Ctrl+v* to paste the figure into the document. You can also save your figure as a structural equation model path diagram (`.stsem`). In this format, you can open it whenever you want to use it again. You should save in this format anytime you create a model you may want to use later. You may also save it as any of several other formats, including as a portable document file (`.pdf`), a picture (`.jpg`), or an encapsulated postscript file (`.eps`).

A.5 Fitting a structural equation model

Let us fit this model. Click on the **Estimate** button, , on the top toolbar, and then click on **OK** in the dialog box that opens. Stata will automatically select the first observed indicator variable for each latent variable as the reference indicator and will make its unstandardized loading fixed at 1.0.[6]

---

6. Our example does not have any observed exogenous variables; however, if we have two or more of them, we do not need to draw in the correlations among them because Stata assumes they will be correlated.

The results for our unstandardized model are shown in figure A.8.

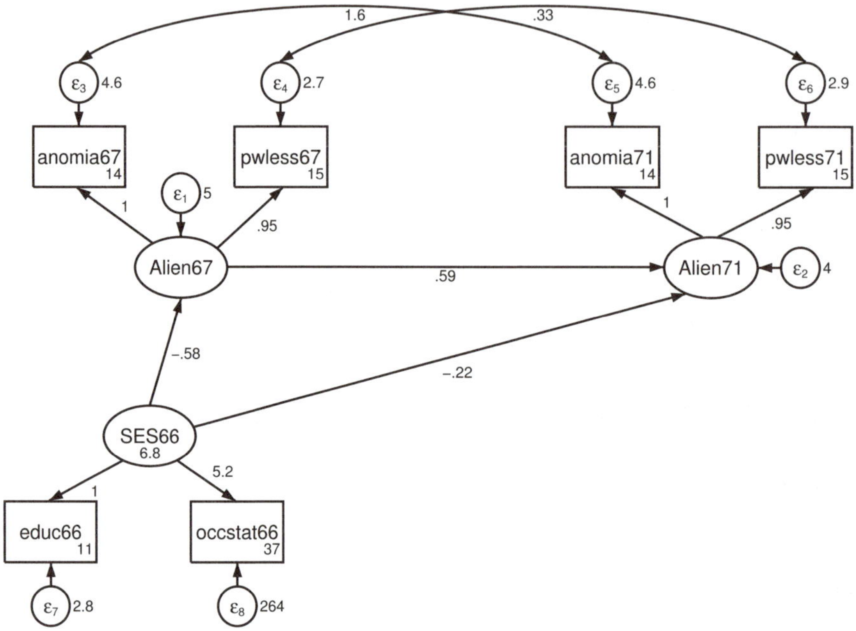

Figure A.8. Unstandardized solution

We can verify that Stata fixed the first indicator of each latent variable as a reference indicator, fixing its unstandardized loading at 1.0. And we can also see that because we constrained the loadings from `Alien67` → `pwless67` and `Alien71` → `pwless71` to be equal, they are both showing the same value, 0.95.

To obtain a standardized solution, click on **View** > **Standardized Estimates**. The standardized results look pretty good, but the figure has some information that could confuse readers. Most published structural equation models do not include the variance of the errors in the observed variables or the intercepts for each of the observed variables. We would not want to include the variance of `SES66` in the oval. We also would want all our values to be consistent, say, appearing to two decimal places.

It is pretty easy to change our output to meet our preferences. Click on **Settings** > **Connections** > **Paths...**. Here click on the tab for **Results**. The first result shows it will report the parameter; you can click on that to see the many other options available. In the lower left below *Appearance of results*, click on **Result 1...**. Among the many options available, we will change the format. In Stata, we specify a format of `%5.2f` to have the parameters reported in a fixed font with five spaces for the maximum and two decimal places. Let us make that change here.

Next click on **Settings** > **Variables** > **All Observed...**, and select the tab marked **Results**. We do not have any exogenous observed variables, but we do have six endogenous observed variables. Because we do not want the intercepts reported for these variables, click on *Intercept* and change it to None. Next click on **Settings** > **Variables** > **All Latent...**, and select the **Results** tab. Here for both exogenous and endogenous variables, change what is reported to None. Finally, click on **Settings** > **Variables** > **All Error...**, and select the **Results** tab. Change the reported results to None.

With our preferences implemented, our figure now looks like figure A.9.

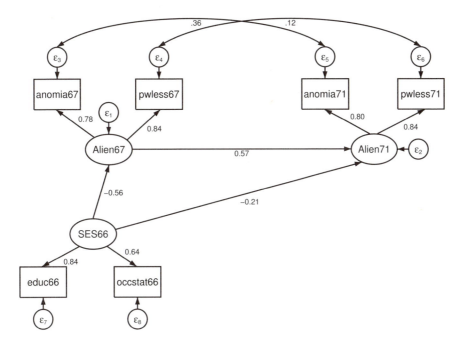

Figure A.9. Standardized results

Stata has options for what it reports along with each of the paths (p-values, z-values, confidence intervals). There is rarely room for this much information in a figure. It is typical to report one asterisk for $p < 0.05$, two asterisks for $p < 0.01$, and three asterisks for $p < 0.001$. To add this reporting in the SEM Builder, we need to **Add Text**. Click on the **Add Text** tool on the drawing toolbar, click on the figure about where you want the asterisks, enter *** in the dialog box that opens, and then click on **OK**. You can always move these text boxes around so that they are properly located. Figure A.3 has asterisks that were added in this way.

A.6   Postestimation commands

The **Estimation** menu provides access to a wide variety of postestimation commands that can be run from within the SEM Builder. Select **Estimation > Goodness of fit > Overall goodness of fit**. This opens a menu that has many options for *Reports and statistics*. Goodness-of-fit statistics (gof) will be selected. For *Statistics to be displayed*, select All of the above. Now our model has a chi-squared test, the root mean squared error of approximation, the comparative fit index, and several other statistics. These do not appear on your figure, but they are in the Results window. Another option under **Estimation > Goodness of fit** is **Equation-level goodness of fit**. This will give you the R^2 for each endogenous variable. Run this to find out that our model explains 31.8% of the variance in Alien67 and 49.8% of the variance in Alien71.

Our model fits the data very well, but if you want modification indices, click on **Estimation > Tests and CIs > Modification indices**. This can help us diagnose problems in a poorly fitting model. In many models, we want indirect effects. The only indirect effect we might want to estimate in our model is SES66 → Alien67 → Alien71. To compute this indirect effect, click on **Estimation > Tests and CIs > Direct and indirect effects**. Check *Do not display effects with no paths*, *Report standardized effects*, *Do not display direct effects* (we already have them), and *Suppress blank lines*. This postestimation command reports that SES66 has a standardized indirect effect of -0.32 on Alien71, and the unstandardized indirect effect is significant: $z = -8.53$, $p < 0.001$. You will find several other postestimation commands in the **Estimation** menu as well.

You may want to add a block of text to the figure that includes information such as the chi-squared value, its level of significance, etc. Suppose you want to add a block of text like the hypothetical one in figure A.10.

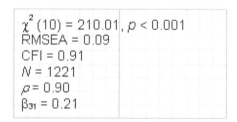

Figure A.10. Text block with Greek letters

Click on the **Add Text** tool, and then click where you want the text to start. In the dialog box that opens, enter the following:

```
{&chi}{sup:2} (10) = 210.01, {it:p} < 0.001
RMSEA = 0.09
CFI = 0.91
{it:N} = 1221
{it:{&rho}} = 0.90
{&beta}{sub:31} = 0.21
```

This looks complicated at first glance, but it is really straightforward. When you want to enter a Greek letter, include an ampersand (**&**) as the first character. Thus to enter a χ, type **&chi** and to enter a ρ, type **&rho**. Notice above that these Greek letters appear inside curly braces. If you do not use curly braces very often in your work, you will probably mistakenly use [] or ( ) (this happens to me, at least). To get a subscript, use **sub:** followed by the subscript number (above, we have **sub:31**, again surrounded by curly braces); for a superscript, use **sup:** in the same manner. To enter an italicized letter, enter **it:** followed by the letter to italicize, again enclosed in curly braces. For a combination, such as an italicized Greek ρ, be sure to use balanced curly braces. In the example above, you see the following combination: **{it:{&rho}} = 0.90**.

If you cannot remember these ways to enter special characters, you can always use Stata's help system. You access this help menu by typing **help sg__tags** (notice the two underscores between **sg** and **tags**).

A.7 Clearing preferences and restoring the defaults

When you use the SEM Builder to create a complex figure, you may need to change several default preferences on such things as the size of rectangles, how far apart they are, etc. The SEM Builder will remember these changes the next time you use it. The defaults you changed may or may not be appropriate for a new figure. You can change settings one at a time as needed, or you can start fresh by restoring the original defaults. To restore the SEM Builder to its default settings, open a new SEM Builder window, and click on **Settings > Settings Defaults**.

The best way to develop skill using the SEM Builder is simply to use it to draw figures of varying complexity. You cannot hurt Stata by making a mistake, and restoring the original settings ensures you can always get a fresh start.

B   Entering data from summary statistics

Many articles report enough information for you to analyze their data. Path analysis and standard structural equation models only need a covariance matrix or a correlation matrix along with the standard deviations. Some models, such as multiple-group models or growth curve models, also need the means. Given just these summary statistics, Stata can fit the same model and get the same results for all the parameters estimates, standard errors, and measures of goodness of fit. You can use the summary statistics to modify the original model in ways you feel are more appropriate, such as by adding a path, correlating a pair of error terms, or creating a nonrecursive model. This capability is extremely useful when you want to reanalyze published results. It also might save some time if you had an unusually large dataset, because Stata does not need to read and analyze the large raw dataset.

Be warned, however, that the results produced by sem are only as good as the summary statistics you enter. Many journals, such as those following *American Psychological Association* guidelines, only report two decimal places. This arbitrary restraint is unfortunate because it gives you less precision in your results than would be possible with more decimal places.

In this example, we will use hypothetical data. Suppose you read a study of the executive functioning of 6-year-old children. Children who have strong working memories, can exercise inhibitory control (wait for a better but less immediate advantage), and have good cognitive skills are said to have strong executive functioning. This strong executive functioning leads to better scores on math achievement. Suppose the study combines three items to measure each of these concepts to obtain a single score on a nine-item executive function scale. The study also uses math achievement when the child is 8 years old to show that executive functioning predicts academic performance. The original model is shown in figure B.1.

After reading this article, you decide that it is better to estimate the individual effects of the three components of executive functioning—working memory, inhibitory control, and cognitive skills—rather than treating the three as a single scale. You think that treating these as three separate variables will do a better job of predicting math performance. You believe that each of these variables may have differential importance in predicting math achievement.

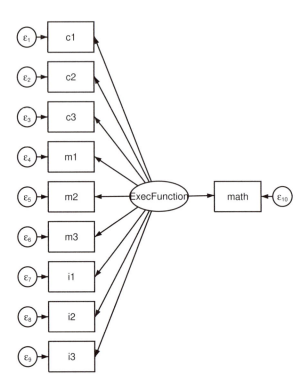

Figure B.1. Original model correlating executive functioning and math achievement

You may have noticed that `ExecFunction` is a long name for a variable, and it does not fit into the default width of the latent variable oval. Use the **Select** tool to click on the `ExecFunction` oval. On the top toolbar, click on **Properties...** to open the *Variable properties* dialog box. In the **Appearance** tab, click on *Customize appearance for selected variables* and then *Set custom appearance*. In the menu that opens, change the size of the oval from `0.65 inch` to `0.75 inch`.

Your model, which is shown in figure B.2, uses the three components of executive functions separately to predict math achievement.

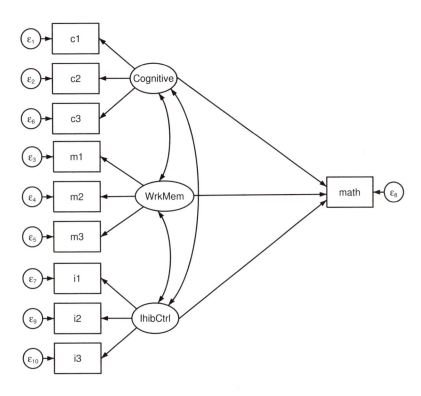

Figure B.2. Your model of cognitive ability, working memory, and inhibitory control to predict math achievement

We use the ssd command (summary statistics data) to enter the original reported summary statistics. First, we use `ssd init c1 c2 c3 m1 m2 m3 i1 i2 i3 math` to provide names for our variables. Next we use `ssd set observations 500` to set the number of observations in the dataset (here we assume there were $N = 500$ children in the study). We next enter the standard deviations for each variable with the `ssd set sd` command. If you want to just do a standardized analysis or you do not know the standard deviations, you could just enter `1.0` as the standard deviation for each variable, which would make all your results standardized. We enter the correlation matrix with `ssd set correlations` to enter everything on the diagonal and below the diagonal. This looks nicer if we change the delimiter temporarily from a carriage return to a semicolon; this way, we do not need to put the three slashes (///) at the end of each line. We change the delimiter back to a carriage return after entering the matrix.

```
. clear

. ssd init c1 c2 c3 m1 m2 m3 i1 i2 i3 math

. ssd set observations 500

. ssd set sd 1.5 1.7 1.6 2.0 1.8 1.8 1.9 1.8 1.7 5.6

. #delimit ;

. ssd set correlations
  1.0 \
  .60 1.0 \
  .50 .55 1.0 \
  .20 .25 .22 1.0 \
  .25 .28 .29 .70 1.0 \
  .22 .25 .24 .60 .60 1.0 \
  .33 .30 .29 .34 .33 .29 1.0\
  .31 .31 .32 .33 .29 .30 .50 1.0\
  .30 .29 .29 .29 .30 .30 .61 .59 1.0\
  .40 .35 .37 .25 .28 .27 .15 .22 .18 1.0 ;

. #delimit cr
```

The results are reported as follows:

```
Status:
                              observations: set
                                     means: unset
                           variances or sd: set
               covariances or correlations: set
```

The means are not set because we did not enter any. We do not need the means to produce an equivalent dataset for our analysis, but we would for other analyses. At this point, it is wise to save this dataset; we would not want to have to reenter all this information. We save it, reopen it, and then run `ssd list` to see what is in the dataset.

```
. save ssdmath.dta
file ssdmath.dta saved

. clear

. use ssdmath

. ssd list

Observations = 500

Means undefined; assumed to be 0

Standard deviations:
   c1    c2    c3    m1    m2    m3    i1    i2    i3  math
  1.5   1.7   1.6     2   1.8   1.8   1.9   1.8   1.7   5.6

Correlations:
   c1    c2    c3    m1    m2    m3    i1    i2    i3  math
    1
   .6     1
   .5   .55     1
   .2   .25   .22     1
  .25   .28   .29    .7     1
  .22   .25   .24    .6    .6     1
  .33    .3   .29   .34   .33   .29     1
  .31   .31   .32   .33   .29    .3    .5     1
   .3   .29   .29   .29    .3    .3   .61   .59     1
   .4   .35   .37   .25   .28   .27   .15   .22   .18     1
```

Everything looks fine, so we are ready to fit our structural equation models. First, we reproduce the model in the original article:

```
. sem (ExecFunction -> c1 c2 c3 m1 m2 m3 i1 i2 i3)
>       (math <- ExecFunction), standardized
 (output omitted)
Structural equation model                      Number of obs      =      500
Estimation method  = ml
Log likelihood     = -9864.5372

 ( 1)   [c1]ExecFunction = 1
```

Standardized	Coef.	OIM Std. Err.	z	P>|z|	[95% Conf. Interval]	
Measurement						
c1 <-						
ExecFunct~n	.5572495	.0381339	14.61	0.000	.4825084	.6319905
c2 <-						
ExecFunct~n	.5743078	.0370245	15.51	0.000	.5017411	.6468744
c3 <-						
ExecFunct~n	.5547813	.0374813	14.80	0.000	.4813193	.6282433
m1 <-						
ExecFunct~n	.6337135	.0363397	17.44	0.000	.562489	.7049381
m2 <-						
ExecFunct~n	.6532323	.0351239	18.60	0.000	.5843908	.7220738
m3 <-						
ExecFunct~n	.6056621	.0363445	16.66	0.000	.5344282	.6768959
i1 <-						
ExecFunct~n	.6087451	.0343895	17.70	0.000	.541343	.6761472
i2 <-						
ExecFunct~n	.606511	.0345234	17.57	0.000	.5388463	.6741757
i3 <-						
ExecFunct~n	.6072909	.035136	17.28	0.000	.5384256	.6761561
math <-						
ExecFunct~n	.4460205	.0406677	10.97	0.000	.3663133	.5257278
var(e.c1)	.689473	.0425002			.6110094	.7780127
var(e.c2)	.6701706	.0425269			.5917943	.7589268
var(e.c3)	.6922177	.0415878			.6153231	.7787215
var(e.m1)	.5984072	.0460579			.5146143	.6958437
var(e.m2)	.5732875	.0458881			.4900486	.6706653
var(e.m3)	.6331735	.0440249			.5525075	.7256167
var(e.i1)	.6294294	.0418688			.5524923	.7170804
var(e.i2)	.6321444	.0418777			.555171	.7197901
var(e.i3)	.6311978	.0426755			.5528604	.7206353
var(e.math)	.8010657	.0362773			.7330277	.8754188
var(ExecFun~n)	1	.			.	.

```
LR test of model vs. saturated: chi2(35)  =      685.17, Prob > chi2 = 0.0000
```

```
. estat eqgof
```

Equation-level goodness of fit

depvars	fitted	Variance predicted	residual	R-squared	mc	mc2
observed						
c1	2.2455	.6972883	1.548212	.310527	.5572495	.310527
c2	2.88422	.9513006	1.932919	.3298294	.5743078	.3298294
c3	2.55488	.7863469	1.768533	.3077823	.5547813	.3077823
m1	3.992	1.603159	2.388841	.4015928	.6337135	.4015928
m2	3.23352	1.379783	1.853737	.4267125	.6532323	.4267125
m3	3.23352	1.186141	2.047379	.3668265	.6056621	.3668265
i1	3.60278	1.335084	2.267696	.3705706	.6087451	.3705706
i2	3.23352	1.189468	2.044052	.3678556	.606511	.3678556
i3	2.88422	1.063707	1.820513	.3688022	.6072909	.3688022
math	31.29728	6.226103	25.07118	.1989343	.4460205	.1989343
overall				.8433886		

```
mc  = correlation between depvar and its prediction
mc2 = mc^2 is the Bentler-Raykov squared multiple correlation coefficient

. estat gof, stats(all)
```

Fit statistic	Value	Description
Likelihood ratio		
chi2_ms(35)	685.170	model vs. saturated
p > chi2	0.000	
chi2_bs(45)	1893.324	baseline vs. saturated
p > chi2	0.000	
Population error		
RMSEA	0.193	Root mean squared error of approximation
90% CI, lower bound	0.180	
upper bound	0.205	
pclose	0.000	Probability RMSEA <= 0.05
Information criteria		
AIC	19769.074	Akaike's information criterion
BIC	19853.366	Bayesian information criterion
Baseline comparison		
CFI	0.648	Comparative fit index
TLI	0.548	Tucker-Lewis index
Size of residuals		
SRMR	0.114	Standardized root mean squared residual
CD	0.843	Coefficient of determination

These are standardized results, so we get a standardized loading for each indicator of the single construct, executive functioning. The cognitive items have loadings of 0.55–0.57; the memory items have loadings of 0.61–0.63; and each of the inhibitory control items has a loading of 0.61. All of this seems reasonable.

The `estat eqgof` command gives us the R^2 for each endogenous variable. We are most interested in the $R^2 = 0.20$ for `math` because that is our outcome variable. Thus we can say that executive functioning explains 20% of the variance in math performance.

The `estat gof, stats(all)` command provides additional information to evaluate the original model. The $\chi^2(35) = 685.17$, $p < 0.001$, the root mean squared error of approximation (RMSEA) $= 0.19$, and the comparative fit index (CFI) $= 0.65$. All these values say this is a lousy fit to the data. The original model does have some predictive power, enabling us to predict which children are more likely to do better or worse at math, but the model does not fit all the relationships in the data. The original model provides a poor fit.

What about our model? Remember that we kept the three dimensions of executive functioning separate. Here is our set of commands:

```
. sem (C -> c1 c2 c3)
>     (M -> m1 m2 m3)
>     (I -> i1 i2 i3)
>     (math <- C M I),
>     standardized
  (output omitted)
Structural equation model              Number of obs      =         500
Estimation method  = ml
Log likelihood     = -9541.8838
  ( 1)   [c1]C = 1
  ( 2)   [m1]M = 1
  ( 3)   [i1]I = 1
```

Standardized		Coef.	OIM Std. Err.	z	P>|z|	[95% Conf. Interval]	
Measurement							
c1 <-							
	C	.7569368	.0271735	27.86	0.000	.7036777	.8101958
c2 <-							
	C	.776716	.0263183	29.51	0.000	.725133	.828299
c3 <-							
	C	.6966369	.0299534	23.26	0.000	.6379293	.7553445
m1 <-							
	M	.8298021	.0212868	38.98	0.000	.7880807	.8715235
m2 <-							
	M	.837679	.0210442	39.81	0.000	.7964332	.8789247
m3 <-							
	M	.7248122	.026242	27.62	0.000	.6733788	.7762455
i1 <-							
	I	.7431854	.0275673	26.96	0.000	.6891545	.7972162

i2 <-						
I	.7165186	.0287439	24.93	0.000	.6601817	.7728555

i3 <-						
I	.8070029	.0250413	32.23	0.000	.7579229	.8560828

math <-						
C	.4799614	.0546839	8.78	0.000	.3727828	.5871399
M	.1890396	.0532602	3.55	0.000	.0846514	.2934277
I	-.1107696	.0622383	-1.78	0.075	-.2327545	.0112153

var(e.c1)	.4270467	.0411372			.3535734	.515788
var(e.c2)	.3967122	.0408837			.3241558	.4855091
var(e.c3)	.514697	.0417333			.4390697	.6033507
var(e.m1)	.3114285	.0353277			.2493447	.3889704
var(e.m2)	.298294	.0352565			.2366125	.3760548
var(e.m3)	.4746473	.038041			.4056494	.5553813
var(e.i1)	.4476755	.0409752			.3741571	.5356397
var(e.i2)	.4866011	.041191			.41221	.5744176
var(e.i3)	.3487464	.0404167			.2778835	.43768
var(e.math)	.7251795	.0400949			.6507032	.8081801
var(C)	1	.			.	.
var(M)	1	.			.	.
var(I)	1	.			.	.

cov(C,M)	.4063289	.0475814	8.54	0.000	.313071	.4995868
cov(C,I)	.5305448	.0444618	11.93	0.000	.4434013	.6176883
cov(M,I)	.4983121	.0437026	11.40	0.000	.4126565	.5839677

LR test of model vs. saturated: chi2(30) =   39.86, Prob > chi2 = 0.1075

. estat eqgof

Equation-level goodness of fit

depvars	fitted	Variance predicted	residual	R-squared	mc	mc2
observed						
c1	2.2455	1.286567	.9589334	.5729533	.7569368	.5729533
c2	2.88422	1.740015	1.144205	.6032878	.776716	.6032878
c3	2.55488	1.239891	1.314989	.485303	.6966369	.485303
m1	3.992	2.748778	1.243222	.6885715	.8298021	.6885715
m2	3.23352	2.26898	.9645395	.701706	.837679	.701706
m3	3.23352	1.698738	1.534782	.5253527	.7248122	.5253527
i1	3.60278	1.989904	1.612876	.5523245	.7431854	.5523245
i2	3.23352	1.660086	1.573434	.5133989	.7165186	.5133989
i3	2.88422	1.878359	1.005861	.6512536	.8070029	.6512536
math	31.29728	8.601133	22.69615	.2748205	.5242332	.2748205
overall				.9913729		

mc = correlation between depvar and its prediction
mc2 = mc^2 is the Bentler-Raykov squared multiple correlation coefficient

```
. estat gof, stats(all)
```

Fit statistic	Value	Description
Likelihood ratio		
chi2_ms(30)	39.863	model vs. saturated
p > chi2	0.108	
chi2_bs(45)	1893.324	baseline vs. saturated
p > chi2	0.000	
Population error		
RMSEA	0.026	Root mean squared error of approximation
90% CI, lower bound	0.000	
upper bound	0.045	
pclose	0.984	Probability RMSEA <= 0.05
Information criteria		
AIC	19133.768	Akaike's information criterion
BIC	19239.133	Bayesian information criterion
Baseline comparison		
CFI	0.995	Comparative fit index
TLI	0.992	Tucker-Lewis index
Size of residuals		
SRMR	0.025	Standardized root mean squared residual
CD	0.991	Coefficient of determination

Have we improved on the original publication? First, we can examine the loadings. We see that the three indicators of cognitive ability have loadings between 0.70 and 0.78, the three indicators of working memory have loadings between 0.72 and 0.84, and the three indicators of inhibitory control have loadings between 0.72 and 0.81. This is impressive!

We see that the standardized paths from the three components of executive functioning have different effects on math performance. The cognitive ability has $\beta = 0.48$, $z = 8.78$, $p < 0.001$. The working memory also has a significant effect, although it is much weaker with $\beta = 0.19$, $z = 3.55$, $p < 0.001$. However, inhibitory control has a negative beta: $\beta = -0.11$, $z = -1.78$, $p - 0.08$. Disentangling these three components shows that they have quite different effects on math achievement (again, we are using hypothetical data). It makes no sense to combine them into a single score.

The R^2 for our model for `math` is 0.28, and this is an improvement over the $R^2 = 0.20$ for the original model. How well does our model fit the data compared with the original model? Our model has $\chi^2(30) = 39.86$, $p = 0.11$, RMSEA $= 0.03$, and CFI $= 1.00$. This is a dramatic improvement when compared with the fit of the original model.

These are hypothetical data and the original article is imaginary. However, we have illustrated how you can replicate an original study and then make changes you feel are important. You can then show whether your idea has more merit than the original idea. This is a powerful capability of Stata, and you should take advantage of it. The use of summary statistics can also help when you have a huge dataset with 100,000s observations; all those observations can be reduced into a very small `ssd` dataset.

Of course, there are a few pitfalls of this approach.

- Many authors do not report the necessary summary statistics, and some journals may even discourage this to save journal space. When the summary statistics are not published, you can sometimes request them from the authors.

- When the appropriate summary statistics are reported, they often are provided with just two or three decimal places, so when you try to reproduce what the authors did, you may get slightly different results.

- We are using structural equation modeling. Many researchers still rely on ordinary least-squares regression, and they were not able to isolate the estimated measurement error from the structural model.

- In analyzing the summary statistics, we are acting as if there were no missing values. The authors of the original study may have reported a listwise correlation matrix or a pairwise correlation matrix, and both of these are ineffective ways of handling missing values.

Despite these limitations, we should be able to fairly closely approximate the results reported in the original study, and we can now use many of the capabilities of structural equation modeling to improve on the original model.

References

Acock, A. C. 2012a. *A Gentle Introduction to Stata*. Revised 3rd ed. College Station, TX: Stata Press.

———. 2012b. What to do about missing values. In *APA Handbook of Research Methods in Psychology, Volume 3: Data Analysis and Research Publication*, ed. H. Cooper, 30–54. New York: American Psychological Association.

Bentler, P. M., and T. Raykov. 2000. On measures of explained variance in nonrecursive structural equation models. *Journal of Applied Psychology* 85: 125–131.

Bollen, K. A., and J. Pearl. 2013. Eight myths about causality and structural equation models. In *Handbook of Causal Analysis for Social Research*, ed. S. L. Morgan, 301–328. Dordrecht: Springer.

Brown, T. A. 2006. *Confirmatory Factor Analysis for Applied Research*. New York: Guilford Press.

Costello, A. B., and J. W. Osborne. 2005. Best practices in exploratory factor analysis: Four recommendations for getting the most from your analysis. *Practical Assessment, Research & Evaluation* 10: 173–178.

Duncan, O. D., A. O. Haller, and A. Portes. 1968. Peer influences on aspirations: A reinterpretation. *American Journal of Sociology* 74: 119–137.

Fabrigar, L. R., D. T. Wegener, R. C. Maccallum, and E. J. Strahan. 1999. Evaluating the use of exploratory factor analysis in psychological research. *Psychological Methods* 4: 272–299.

Graham, J. W. 2009. Missing data analysis: Making it work in the real world. *Annual Review of Psychology* 60: 549–576.

Grimm, K. J., N. Ram, and F. Hamagami. 2011. Nonlinear growth curves in developmental research. *Child Development* 82: 1357–1371.

Halpern, J. Y., and J. Pearl. 2005. Causes and explanations: A structural-model approach—Part I: Causes. *British Journal of Philosophy of Science* 56: 843–887.

Hedström, P., and R. Swedberg, ed. 1998. *Social Mechanisms: An Analytical Approach to Social Theory*. Cambridge: Cambridge University Press.

Hu, L., and P. M. Bentler. 1999. Cutoff criteria for fit indexes in covariance structure analysis: Conventional criteria versus new alternatives. *Structural Equation Modeling* 6: 1–55.

Kline, P. 2000. *Handbook of Psychological Testing*. 2nd ed. London: Routledge.

Little, T. D., W. A. Cunningham, G. Shahar, and K. F. Widaman. 2002. To parcel or not to parcel: Exploring the question, weighing the merits. *Structural Equation Modeling* 9: 151–173.

Marsh, H. W., and D. Hocevar. 1985. Application of confirmatory factor analysis to the study of self-concept: First- and higher-order factor models and their invariance across groups. *Psychological Bulletin* 97: 562–582.

McClelland, M. M., A. C. Acock, A. Piccinin, S. A. Rhea, and M. C. Stallings. 2013. Relations between preschool attention span–persistence and age 25 educational outcomes. *Early Childhood Research Quarterly* 28: 314–324.

Medeiros, R. 2012. Personal communication.

Mitchell, M. N. 2012. *A Visual Guide to Stata Graphics*. 3rd ed. College Station, TX: Stata Press.

Raykov, T. 1997a. Estimation of composite reliability for congeneric measures. *Applied Psychological Measurement* 21: 173–184.

———. 1997b. Scale reliability, Cronbach's coefficient alpha, and violations of essential tau-equivalence with fixed congeneric components. *Multivariate Behavioral Research* 32: 329–353.

Schafer, J. L., and J. W. Graham. 2002. Missing data: Our view of the state of the art. *Psychological Methods* 7: 147–177.

Schreiber, J. B., A. Nora, F. K. Stage, E. A. Barlow, and J. King. 2006. Reporting structural equation modeling and confirmatory factor analysis results: A review. *Journal of Educational Research* 99: 323–337.

Shadish, W. R., T. D. Cook, and D. T. Campbell. 2002. *Experimental and Quasi-Experimental Designs for Generalized Causal Inference*. Boston: Houghton Mifflin.

StataCorp. 2013. *Stata 13 Structural Equation Modeling Reference Manual*. College Station, TX: Stata Press.

U.S. Department of Health and Human Services. 2010. Head Start Impact Study: Final Report. Washington DC: Administration for Children and Families: Office of Planning, Research, and Evaluation. http://www.acf.hhs.gov/node/8375.

Wheaton, B., B. Muthén, D. F. Alwin, and G. F. Summers. 1977. Assessing reliability and stability in panel models. In *Sociological Methodology 1977*, ed. D. R. Heise, 84–136. San Francisco: Jossey-Bass.

Author index

Subject index